Algebraic Aspects of the Advanced Encryption Standard

Algebraic Aspects of the Advanced Encryption Standard

by

Carlos Cid
Royal Holloway, University of London
United Kingdom

Sean Murphy
Royal Holloway, University of London
United Kingdom

Matthew Robshaw
France Telecom R&D
France

 Springer

Carlos Cid
Information Security Group
Royal Holloway
University of London
Egham, Surrey TW20 0EX
United Kingdom
email: carlos.cid@rhul.ac.uk

Sean Murphy
Information Security Group
Royal Holloway
University of London
Egham, Surrey TW20 0EX
United Kingdom
email: s.murphy@rhul.ac.uk

Matthew Robshaw
France Telecom Research and Development
38-40 rue du General-Leclerc
92794 Issy les Moulineaux, France
email: matt.robshaw@orange-ft.com

by Carlos Cid, Sean Murphy and Matthew Robshaw

e-ISBN-10: 0-387-36842-6

ISBN-13: 978-1-4419-3729-2

e-ISBN-13: 978-0-387-36842-9

Printed on acid-free paper.

9 8 7 6 5 4 3 2 1

springer.com

Contents

Preface

It is now more than five years since the Belgian block cipher *Rijndael* was chosen as the *Advanced Encryption Standard (AES)*. Joan Daemen and Vincent Rijmen used algebraic techniques to provide an unparalleled level of assurance against many standard statistical cryptanalytic techniques. The cipher is a fitting tribute to their distinctive approach to cipher design. Since the publication of the AES, however, the very same algebraic structures have been the subject of increasing cryptanalytic attention and this monograph has been written to summarise current research. We hope that this work will be of interest to both cryptographers and algebraists and will stimulate future research.

During the writing of this monograph we have found reasons to thank many people. We are especially grateful to the British *Engineering and Physical Sciences Research Council* (EPSRC) for their funding of the research project *Security Analysis of the Advanced Encryption System* (Grant GR/S42637), and to Susan Lagerstrom-Fife and Sharon Palleschi at Springer. We would also like to thank Claus Diem, Maura Paterson, and Ludovic Perret for their valuable comments. Finally, the support of our families at home and our colleagues at work has been invaluable and particularly appreciated.

May 2006   *Carlos Cid, Sean Murphy, and Matt Robshaw*

Preface

Chapter 1

INTRODUCTION TO THE AES

In January 1997, the U.S. National Institute of Standards and Technology (NIST) announced the impending development of an *Advanced Encryption Standard* (*AES*). The intention was that it would:

> ...specify an unclassified, publicly disclosed encryption algorithm capable of protecting sensitive government information well into the next century [94].

To achieve this goal NIST established an open competition. Fifteen block ciphers from around the world were submitted as candidates, with the initial field being narrowed to a set of five in the first round. From these five candidates the block cipher Rijndael [37], designed by Belgian cryptographers Joan Daemen and Vincent Rijmen, was chosen as the AES. The AES was finally published as the Federal Information Processing Standard *FIPS 197* [95] in November 2001.

1. Background

The AES is a block cipher and therefore encrypts and decrypts blocks of data according to a secret key. The AES is intended to replace the *Data Encryption Standard* (*DES*) [92], and gradually to replace *Triple-DES* [96]. The DES and Triple–DES have been remarkably successful block ciphers and have been used in millions of systems around the world. Even after thirty years of analysis, the most practical attack remains exhaustive key search. However, the restricted key and block size of the DES, along with its relatively poor performance in software, make a replacement inevitable.

The AES has a very different structure to the DES. Whilst the DES is said to be a *Feistel* cipher [50], the AES is said to be a *substitution–permutation* (SP-)*network* [113]. Both the AES and the DES are *iterated* ciphers, which means that a certain sequence of computations, constituting a *round*, is repeated a specified number of times. The operations used within the AES are byte-oriented and the cipher offers good performance in hardware, limited byte-oriented processors, and modern 32-bit and 64-bit machines. By contrast, the operations used in the DES are fundamentally bit operations. As a consequence the DES offers outstanding performance in hardware but offers generally poor performance in most software environments.

The AES and the DES are closely related in terms of their underlying design philosophies. Both rely heavily on the ideas of Shannon [113] and the concepts of *diffusion* and *confusion*. Whilst most block ciphers follow these principles, few do so as clearly as the AES and the DES. The aim of diffusion is to spread the influence of all parts of the inputs to a block cipher, namely the *plaintext* and the *key*, to all parts of the output, the *ciphertext*. Diffusion is provided in both the AES and DES by the use of permutations. The aim of confusion is to make the relationship between plaintext, ciphertext, and key complicated. In both the AES and DES, confusion is provided by very carefully chosen *substitution* or *S-boxes*. These make local substitutions of small sub-blocks of data and these local changes are then spread by the diffusion transformations.

From the start of the AES selection process there was a significant level of support for Rijndael. Whilst Rijndael had a relatively unfamiliar design, there had been two immediate predecessors. The block cipher SHARK was proposed in 1996 [108] and featured all the components later used in Rijndael. Some changes to SHARK would lead a year later to the block cipher SQUARE [36]. The proposal for SQUARE also includes a discussion of the *square attack* for this type of block cipher.

The initial popular support for Rijndael increased in the second round of the AES selection process. The significance of the design strategy for Rijndael was clear on two counts. Firstly, the transparent design permitted quick and accurate security estimates to be made for Rijndael against standard attacks. Secondly, the byte-wise design helped give Rijndael a versatile performance characteristic. Rijndael was generally perceived to be a worthy selection as the AES. As a brief review of the state of the AES five years after its selection concludes:

> ... there have been few cryptanalytic advances despite the efforts of many researchers. The most promising new approach to AES cryptanalysis remains speculative, while the most effective attack against versions of the AES with fewer rounds is older than the AES itself [42].

This "new approach" is algebraic in nature and is the subject of this monograph.

2. Algebraic Perspectives

With hindsight it is a little surprising that the algebraic properties of Rijndael were not discussed more during the AES selection process. Many observers noted that the structured design of Rijndael might have interesting side-effects, most obviously the ongoing development of dedicated attacks such as the *square* [51] and *bottleneck* [55] attacks. However, the algebraic foundations of Rijndael were not explored in detail.

An early public comment about Rijndael with such an algebraic perspective [88] appeared towards the end of the AES selection process. Other research around this time [52, 67, 111] made similar points. The transparent structure of the AES and its strong algebraic foundations give an interesting framework for analysis. In particular, there are many alternative representations of part or all of the cipher. Some representations provide interesting insights into the interaction of different operations. Others provide implementation benefits, either in terms of security with regards to side-channel cryptanalysis or in terms of performance improvements. However, some representations, such as those that have sought to represent the process of AES encryption as a system of equations, may point the way to future breakthroughs.

Whilst little research was undertaken about the algebraic properties of Rijndael during the AES selection process itself, there has been much research since. One intriguing recent development has been the interplay between the cryptanalysis of symmetric ciphers and that of asymmetric ciphers. Due to the fundamental reliance of asymmetric cryptography on computational algebra and number theory, asymmetric cryptanalysis inevitably focuses on the manipulation of algebraic structures. However, this is a very new area for symmetric cryptanalysis, where recognising statistical patterns of bits has traditionally been the most effective form of cryptanalysis. In fact, the overlap between symmetric and asymmetric cryptanalysis is very specific. Several asymmetric cryptosystems have been proposed that depend for their security on the difficulty of solving a large system of multivariate equations of low degree. Similarly, the process of AES encryption can also be expressed in this way. Thus the difficulty of solving such a multivariate equation system is directly related to the security of the AES.

3. Overview of the Monograph

The purpose of this monograph is to provide both an overview and more background to the algebraic analysis of the AES. In Chapter 2 we give the required mathematical foundations. A description of the AES follows in Chapter 3 along with a description of small scale variants to encourage practical experimentation. Chapter 4 considers structural aspects of the AES and the use of different representations. Chapters 5 and 6 consider how we might represent an AES encryption as a system of equations and the possible methods of solution for such systems.

Although we provide a full bibliography, the sources below are particularly helpful. A thorough overview of mathematical cryptology is given in [80]. Unsurprisingly, the most complete background to the evolution and theoretical underpinnings of the AES comes from the designers [35, 37, 107], most particularly in the book *The Design of Rijndael* [39]. Other surveys of the AES are available. An enjoyable series of articles surveying both the DES and the AES is given in [69–71], and a brief overview of the first five years of the AES is provided by [42]. The ECRYPT European Network of Excellence gives a comprehensive review of the AES in *The State of the Art of AES Cryptanalysis* [43], whilst the AES LOUNGE [44] is an online repository of information about the AES covering many non-algebraic aspects of cryptanalysis of the AES as well as its implementation. The AES is formally described in FIPS 197 [95].

Chapter 2

MATHEMATICAL BACKGROUND

This chapter presents the important mathematical definitions and concepts required in this monograph. They are presented in a logical order, with each definition building on earlier concepts. However, the broad goals of the analysis presented in this monograph should be reasonably clear with only a passing acquaintance of the mathematics in this chapter. For more background and context to this mathematical material, we recommend the following references [23, 33, 57–59, 74, 97].

1.   Groups, Rings, and Fields

Groups, rings, and fields constitute the basic structures of abstract algebra. They are also the basic algebraic structures required for the definition and the algebraic analysis of the AES.

Groups

DEFINITION 2.1 Let G be a non–empty set with a binary operation $\circ\colon G \times G \to G$. We say that $(G, \circ)$ is a *group* if the following conditions hold.

- The operation $\circ$ is associative, that is $(g_1 \circ g_2) \circ g_3 = g_1 \circ (g_2 \circ g_3)$ for all $g_1, g_2, g_3 \in G$.

- There exists an element $e \in G$ such that $e \circ g = g \circ e = g$ for all $g \in G$. This element e is unique and is called the *identity element*.

- For every $g \in G$, there exists a unique element $g^{-1} \in G$ such that $g \circ g^{-1} = g^{-1} \circ g = e$. This element g^{-1} is called the *inverse* of g.

The *order* of a group $(G, \circ)$ is the cardinality of the set G and is often denoted by $|G|$. If the order of $(G, \circ)$ is finite, we say that G is a finite group. Similarly, we say that an element $g \in G$ has finite order if there exists a positive integer m such that $g \circ \ldots \circ g = g^m = e$. In this case, the least such integer m is called the *order* of g and is denoted by $o(g)$, and so the inverse element $g^{-1} = g^{o(g)-1}$. For a finite group G, the order of any element divides the order of the group G.

DEFINITION 2.2 The group $(G, \circ)$ is said to be an *abelian* or *commutative* group if $g \circ g' = g' \circ g$ for all $g, g' \in G$.

The group operation $\circ$ is usually clear from the context. When this is the case, the symbol $\circ$ is omitted and the group $(G, \circ)$ denoted by G.

EXAMPLE 2.3 The set of integers $\mathbb{Z}$ under the operation of addition forms an abelian group. Similarly, if n is a positive integer, the set of integers $\mathbb{Z}_n = \{0, \ldots, n-1\}$ under the operation of addition modulo n forms an abelian group of order n.   □

EXAMPLE 2.4 The set of integers $\mathbb{Z}_p^* = \{1, \ldots, p-1\}$ under the operation of multiplication modulo p forms an abelian group if p is prime.   □

EXAMPLE 2.5 Suppose that G_1 and G_2 are groups, then $G = G_1 \times G_2$ is a group with operation defined as $(g_1, g_2) \circ (g_1', g_2') = (g_1 g_1', g_2 g_2')$. The group G is known as the *direct product* of G_1 and G_2.   □

A non–empty subset $H \subset G$ is called a *subgroup* of G if H is itself a group under the same operation. For a finite group, *Lagrange's Theorem* states that the order of any subgroup divides the order of the group. A subgroup H of G is called a *normal subgroup* of G if $g^{-1}hg \in H$ for all $g \in G$ and $h \in H$. The notation $H < G$ and $H \lhd G$ is used to denote that H is a subgroup of G and that H is a normal subgroup of G respectively. A group that has no non–trivial normal subgroups is called a *simple group*.

If H is a subgroup of G, then the *right coset* of H in G defined by $g \in G$ is the set $Hg = \{hg | h \in H\}$. The set of right cosets, $\{Hg | g \in G\}$, forms a partition of the elements of G. We can also define *left cosets* of H in G in a similar manner. The set of right cosets of H in G and the set of left cosets of H in G have the same cardinality. This cardinality is known as the *index* of H in G and is denoted by $[G : H]$. If H is a normal subgroup of G, then the right coset and left coset defined by any $g \in G$ are identical, and $Hg = gH$ is simply called the *coset* of H in G defined by $g \in G$. In this case, the set of all cosets of H in G forms a group with binary operation $(Hg, Hg') \mapsto Hgg'$ for all $g, g' \in G$. This group is called the *quotient group* of G by H. This group has order $[G : H]$ and is denoted by G/H.

DEFINITION 2.6 Let S be a non–empty subset of G. Then the *group generated by S* is defined as the set of all finite products of form $g_1 \circ \ldots \circ g_k$, where either $g_i \in S$ or $g_i^{-1} \in S$.

The group generated by S is denoted by $\langle S \rangle$ and is the smallest subgroup of G which contains S. If $S = \{g\}$, then the group $\langle S \rangle = \langle g \rangle$ generated by a single element $g \in G$ is called the *cyclic group* generated by g. If g has finite order, then $\langle g \rangle = \{g, g^2, \ldots, g^{o(g)-1}, e\}$.

A permutation of a non–empty set $\mathcal{X}$ is a bijective mapping $\mathcal{X} \to \mathcal{X}$. The set of permutations of $\mathcal{X}$, under the operation of composition, forms a group known as the *symmetric group of $\mathcal{X}$*. We denote this group by $S_{\mathcal{X}}$. If $\mathcal{X}$ is finite with cardinality n, this group is also known as the symmetric group on n elements and is denoted by S_n. The order of the group S_n is $n!$. An element of the group S_n that permutes two elements of $\mathcal{X}$ and leaves the remaining elements fixed is called a *transposition*. An element $g \in S_n$ is said to be an *even* permutation if it can be expressed as a product of an even number of transpositions, otherwise g is said to be an *odd* permutation. The subset of S_n consisting of all even permutations is a normal subgroup of S_n, known as the *alternating group* on n elements and is denoted by A_n. For $n > 1$, the order of A_n is $\frac{1}{2}n!$. Furthermore, A_n is a simple group for $n \neq 4$.

DEFINITION 2.7 Let $\mathcal{X}$ be a non–empty set and G a group. A *group action* of G on $\mathcal{X}$ is a mapping $G \times \mathcal{X} \to \mathcal{X}$, denoted by $(g, x) \mapsto g \cdot x$, such that the following two conditions hold.

- If e is the identity of G, then $e \cdot x = x$ for every $x \in \mathcal{X}$;

- $g \cdot (g' \cdot x) = (gg') \cdot x$ for all $g, g' \in G$ and for all $x \in \mathcal{X}$.

If there is a group action of a group G on a set $\mathcal{X}$, we say that the group G *acts on the set $\mathcal{X}$*. An example of a group action is the action of the symmetric group $S_{\mathcal{X}}$ on the set $\mathcal{X}$ defined by $(g, x) \mapsto g(x)$ for all permutations g of $S_{\mathcal{X}}$ and $x \in \mathcal{X}$.

If G is a group acting on the set $\mathcal{X}$, then the *orbit* of $x \in \mathcal{X}$ is defined to be $\{g \cdot x \mid g \in G\} \subset \mathcal{X}$. The orbits of $\mathcal{X}$ form a partition of $\mathcal{X}$. The *stabilizer* of an element $x \in \mathcal{X}$ is defined to be $G_x = \{g \in G \mid g \cdot x = x\}$ and is a subgroup of G. The number of elements in the orbit of $x \in \mathcal{X}$ is the index $[G : G_x]$. Furthermore, if $Fix(g)$ denotes the number of elements of $\mathcal{X}$ that are fixed by $g \in G$, then the number of orbits of G on $\mathcal{X}$ is

$$\frac{1}{|G|} \sum_{g \in G} Fix(g).$$

If the action of G on $\mathcal{X}$ has only one orbit, then for any pair of elements $x, x' \in \mathcal{X}$ there exists $g \in G$ such that $g \cdot x = x'$. In this case the action of G on $\mathcal{X}$ is said to be *transitive*. Furthermore, if for any pair of m-tuples $(x_1, \ldots, x_m), (x'_1, \ldots, x'_m) \in \mathcal{X}^m$ with distinct entries $(x_i \neq x_j$ and $x'_i \neq x'_j)$ there exists $g \in G$ such that $g \cdot x_i = x'_i$, then the action is said to be *m-transitive*. The action is said to be *sharply m-transitive* if such an element $g \in G$ is unique.

If G acts on a set $\mathcal{X}$, then $\mathcal{Y} \subseteq \mathcal{X}$ is called a *block* of G if for every $g \in G$, we have either $g(\mathcal{Y}) = \mathcal{Y}$ or $g(\mathcal{Y}) \cap \mathcal{Y} = \emptyset$. The group G is said to be *primitive* if it has no non–trivial blocks, and *imprimitive* otherwise.

EXAMPLE 2.8 The symmetric group S_n acting on a set of n elements is a primitive and sharply n-transitive group. The alternating group A_n acting on a set of n elements is a primitive and sharply $(n-2)$-transitive $(n > 2)$ group. □

DEFINITION 2.9 Let $(G, \circ)$ and $(H, \cdot)$ be groups. A mapping $\phi \colon G \to H$ is a (group) *homomorphism* if, for all $g, g' \in G$,

$$\phi(g \circ g') = \phi(g) \cdot \phi(g').$$

An injective homomorphism is called a *monomorphism* and a surjective homomorphism is called an *epimorphism*. A bijective homomorphism $\phi \colon G \to H$ is called an *isomorphism*, and the groups G and H are said to be *isomorphic*, denoted by $G \cong H$. An isomorphism from G to itself is called an *automorphism* of G.

DEFINITION 2.10 If $\phi \colon G \to H$ is a homomorphism and e_H is the identity element of H, then the subset

$$\ker \phi = \{g \in G | \phi(g) = e_H\}$$

of G is called the *kernel* of the homomorphism ϕ.

We note that $\ker \phi$ is a normal subgroup of G and the *First Isomorphism Theorem* states that the quotient group $G/\ker \phi$ is isomorphic to the image of ϕ. Furthermore, any normal subgroup $H \triangleleft G$ is the kernel of the "natural" epimorphism $G \to G/H$ defined by $g \mapsto Hg$.

EXAMPLE 2.11 Let H be the group $(\{-1, 1\}, \times)$, where $\times$ denotes the usual operation of integer multiplication. There exists a homomorphism from the symmetric group S_n onto H that maps every even permutation to 1 and every odd permutation to -1. The kernel of this homomorphism consists of all even permutations and so is the alternating group A_n. Thus the quotient group S_n/A_n is isomorphic to H. □

Isomorphic groups have identical algebraic structure and can be regarded as essentially the *same* algebraic object. Isomorphisms are often useful for solving problems that would otherwise be intractable. Thus obtaining alternative representations using isomorphisms is a common technique for the study and analysis of algebraic structures. We note however that constructing isomorphisms between two algebraic structures, and even constructing the inverse isomorphism of a known isomorphism, can often be a very difficult problem.

EXAMPLE 2.12 Let p be a prime number, and $\mathbb{Z}_{p-1}$ and $\mathbb{Z}_p^*$ denote the groups defined in Examples 2.3 and 2.4 respectively. The group $\mathbb{Z}_{p-1}$ is generated additively by the element $1 \in \mathbb{Z}_{p-1}$, and the group $\mathbb{Z}_p^*$ is generated multiplicatively by some $g \in \mathbb{Z}_p^*$. These groups are isomorphic, and an isomorphism between them can be defined by $m \mapsto g^m$, that is the exponentiation in $\mathbb{Z}_p^*$. The inverse isomorphism is known as the *discrete logarithm*, and the calculation of the discrete logarithm is generally believed to be a hard problem. The difficulty of computing this inverse isomorphism is the foundation of the security of many asymmetric cryptosystems, for example the *Digital Signature Standard* [93]. □

Rings

DEFINITION 2.13 Let R be a non–empty set with two associative binary operations $+, \cdot : R \times R \to R$. We say that $(R, +, \cdot)$ is a *ring* (*with unit*) if the following conditions hold.

- $(R, +)$ is an abelian group.

- The operation $\cdot$ is distributive over $+$, that is for all $r, r', r'' \in R$,

$$r \cdot (r' + r'') = r \cdot r' + r \cdot r'' \text{ and } (r' + r'') \cdot r = r' \cdot r + r'' \cdot r.$$

- There is an element $1 \in R$ such that $1 \cdot r = r \cdot 1 = r$ for all $r \in R$.

The identity element of the group $(R, +)$ is usually denoted by 0 and is called the *zero* of the ring $(R, +, \cdot)$. The element 1 is called the *identity element* of the ring $(R, +, \cdot)$.

DEFINITION 2.14 The ring $(R, +, \cdot)$ is a *commutative ring* if $r \cdot r' = r' \cdot r$ for all $r, r' \in R$, that is the operation $\cdot$ is commutative.

All rings considered in this monograph are commutative rings with unit. As with groups, we often assume that the operations $+$ and $\cdot$ are clear, and we denote the ring $(R, +, \cdot)$ simply by R. We also often denote $r \cdot r'$ simply by rr' for $r, r' \in R$.

A commutative ring R is called an *integral domain* if it contains no *zero–divisors*, that is $rr' \neq 0$ for all $r, r' \in R \setminus \{0\}$. A nonzero element r of a ring R is said to be *invertible* (or a *unit*) if there exists $r^{-1} \in R$ such that $r \cdot r^{-1} = r^{-1} \cdot r = 1$. The set of all invertible elements of R is denoted by R^* and forms a group under multiplication known as the *group of units* of R. If all nonzero elements of a ring R are invertible, then R is called a *division ring* and $R^* = R \setminus \{0\}$.

EXAMPLE 2.15 The set of integers $\mathbb{Z}$ under the operations of integer addition and multiplication forms a commutative ring. Similarly, the set of integers $\mathbb{Z}_n = \{0, \ldots, n-1\}$ under the operations of addition and multiplication modulo n forms a commutative ring. We note that $\mathbb{Z}_n$ is a division ring if and only if n is prime. $\qquad\qquad\square$

DEFINITION 2.16 Let $(R, +, \cdot)$ be a ring and I a non–empty subset of R. We say that I is an *ideal* of R, denoted by $I \lhd R$, if the following conditions hold.

- $(I, +)$ is a subgroup of $(R, +)$.

- For all $x \in I$ and $r \in R$, $x \cdot r \in I$ and $r \cdot x \in I$.

The *coset* of an ideal I in R defined by $r \in R$ is denoted by $I + r$ and defined to be the set $\{s + r | s \in I\}$. The cosets of an ideal $I \lhd R$ form a partition of the ring R. The set of all cosets of I forms a ring with addition and multiplication defined by $(I + r) + (I + r') = I + (r + r')$ and $(I + r)(I + r') = I + rr'$ respectively. This ring is denoted by R/I and is called the *quotient ring* or the *residue class ring modulo I*.

If S is a non–empty subset of R, then the *ideal generated by S* is denoted by $\langle S \rangle$ and consists of all finite sums of the form $\sum r_i s_i$, where $r_i \in R$ and $s_i \in S$. An ideal is said to be a *principal ideal* if it can be generated by one element $r \in R$. An integral domain in which every ideal is a principal ideal is called a *principal ideal domain*.

DEFINITION 2.17 If R and R' are rings, then $\phi \colon R \to R'$ is a (ring) *homomorphism* if the following conditions hold.

- $\phi(r + r') = \phi(r) + \phi(r')$ for all $r, r' \in R$.

- $\phi(r \cdot r') = \phi(r) \cdot \phi(r')$ for all $r, r' \in R$.

Different types of ring homomorphism are defined in a similar manner to group homomorphisms. The *kernel* $\ker \phi = \{r \in R | \phi(r) = 0\}$ of a ring homomorphism $\phi \colon R \to R'$ is an ideal of R. Furthermore, the quotient ring $R/\ker \phi$ is isomorphic to the image of R, and every ideal $I \lhd R$ is the kernel of the "natural" epimorphism $R \to R/I$ defined by $r \mapsto I + r$.

Fields

DEFINITION 2.18 A commutative division ring $\mathbb{F}$ is called a *field*.

Thus a field $\mathbb{F}$ is a ring $(\mathbb{F}, +, \cdot)$ such that both $(\mathbb{F}, +)$ and $(\mathbb{F} \setminus \{0\}, \cdot)$ are commutative groups.

EXAMPLE 2.19 The sets $\mathbb{Q}$ of rational numbers, $\mathbb{R}$ of real numbers, and $\mathbb{C}$ of complex numbers form fields under the usual operations of addition and multiplication. □

EXAMPLE 2.20 The set $\mathbb{Z}_n = \{0, \ldots n - 1\}$ under addition and multiplication modulo an integer n is a field if and only if n is prime (Examples 2.3 and 2.4). □

If $\mathbb{F}$ is a field, we say that $\mathbb{F}$ has *positive characteristic* if there exists a positive integer m such that the m-fold sum $1 + \ldots + 1 = 0$. In this case, the least such integer m is called the *characteristic* of $\mathbb{F}$. If there is no such m, we say that $\mathbb{F}$ has *characteristic zero*. The infinite fields $\mathbb{Q}$, $\mathbb{R}$, and $\mathbb{C}$ all have characteristic zero, whilst the finite field $\mathbb{Z}_p$ has characteristic p. In fact, all finite fields have characteristic p for some prime p. We discuss further aspects of finite fields in Section 2.4.

2. Polynomial Rings

Polynomial rings are a special example of commutative ring that play an important role in the theory of finite fields. The algebraic analysis of the AES makes extensive use of polynomial rings.

Univariate polynomial rings

A *monomial* in the single variable or indeterminate x is the formal expression x^i for some $i \in \mathbb{N}$, that is some non–negative power of x. The *degree* of the monomial x^i is i.

DEFINITION 2.21 A univariate *polynomial* in the variable x over a field $\mathbb{F}$ is a finite linear combination over $\mathbb{F}$ of monomials in x, that is a formal expression of the form

$$c_d x^d + c_{d-1} x^{d-1} + \ldots + c_2 x^2 + c_1 x + c_0,$$

where d is a non–negative integer and $c_d, \ldots, c_0 \in \mathbb{F}$, with $c_d \neq 0$ if $d > 0$.

DEFINITION 2.22 The set of all univariate polynomials in the variable x over a field $\mathbb{F}$ forms a ring under the standard operations of polynomial addition and multiplication. This ring is a principal ideal domain called the *univariate polynomial ring* over $\mathbb{F}$ and is denoted by $\mathbb{F}[x]$.

Let $f(x) \in \mathbb{F}[x]$ be a univariate polynomial. The *degree* of $f(x)$ is the maximum integer d such that $c_d \neq 0$, and is denoted by $\deg(f(x))$. If

$$f(x) = c_d x^d + \ldots + c_1 x + c_0,$$

then the summands $c_i x^i$ ($c_i \neq 0$) are called the *terms* of $f(x)$, and c_i is called the *coefficient* of the monomial x^i. Furthermore, we can define the *leading monomial*, *leading coefficient*, and *leading term* of $f(x)$ as x^d, c_d and $c_d x^d$ respectively. A polynomial $f(x)$ is a *monic* polynomial if its leading coefficient is 1.

The *evaluation* of the polynomial $f(x)$ at $a \in \mathbb{F}$ is defined as the element $\sum_{i=0}^{d} c_i a^i \in \mathbb{F}$ and is denoted by $f(a)$. We say that a is a *root* of $f(x)$ if $f(a) = 0$. A polynomial of degree d has at most d roots in $\mathbb{F}$.

THEOREM 2.23 *Univariate Division Algorithm.* Given $f(x)$ and $g(x) \in \mathbb{F}[x]$, then there exists $q(x), r(x) \in \mathbb{F}[x]$ with $\deg(r(x)) < \deg(g(x))$ such that $f(x) = q(x)g(x) + r(x)$. The univariate polynomial $r(x)$ is known as the *remainder* of the division of $f(x)$ by $g(x)$.

The well-known Euclidean algorithm to find the greatest common divisor of two polynomials is just the repeated application of Theorem 2.23.

EXAMPLE 2.24 Suppose that

$$f(x) = x^6 + x^5 + x^3 + x^2 + x + 1 \text{ and } g(x) = x^4 + x^3 + 1$$

are polynomials in the univariate polynomial ring $\mathbb{Z}_2[x]$. We then have

$$x^6 + x^5 + x^3 + x^2 + x + 1 = x^2(x^4 + x^3 + 1) + (x^3 + x + 1),$$

so $f(x) = q(x)f(x) + r(x)$, where $q(x) = x^2$ and $r(x) = x^3 + x + 1$. $\square$

A polynomial $f(x) \in \mathbb{F}[x]$ of positive degree is said to be *irreducible* in $\mathbb{F}[x]$ if there is no factorisation of the form $f(x) = p(x)q(x)$, where $p(x)$ and $q(x)$ are polynomials of positive degree in $\mathbb{F}[x]$. Every polynomial in $\mathbb{F}[x]$ can be written as the product of monic irreducible polynomials and some constant in $\mathbb{F}$, and this product is unique up to the order of the factors.

EXAMPLE 2.25 Let $f(x)$ be a polynomial in $\mathbb{F}[x]$ of degree d, and $\langle f(x) \rangle$ be the ideal generated by $f(x)$. The elements of the quotient ring $\frac{\mathbb{F}[x]}{\langle f(x) \rangle}$ can be written as polynomials

$$a_{d-1} x^{d-1} + \ldots + a_1 x + a_0$$

in $\mathbb{F}[x]$ of degree less than d. In this representation of the quotient ring, addition is simply polynomial addition. However, multiplication in the quotient ring is defined by applying Theorem 2.23. For two polynomials $g_1(x), g_2(x) \in \mathbb{F}[x]$, we know that there exists $q(x), r(x) \in \mathbb{F}[x]$ such that

$$g_1(x)g_2(x) = q(x)f(x) + r(x),$$

where $deg(r(x)) < deg(f(x)) = d$. In this representation of the quotient ring $\frac{\mathbb{F}[x]}{\langle f(x) \rangle}$, the product of $g_1(x)$ and $g_2(x)$ is $r(x)$. $\qquad\Box$

EXAMPLE 2.26 Let $f(x) = x^5 + x^4 + 1$ be a polynomial in the univariate polynomial ring $\mathbb{Z}_2[x]$. The product of the polynomials $(x^4 + x^3 + x^2 + 1)$ and $(x^4 + x^3 + x + 1)$ satisfies

$$
\begin{aligned}
(x^4 + x^3 + x^2 + 1)(x^4 + x^3 + x + 1) &= x^8 + x^4 + x^3 + x^2 + x + 1 \\
&= (x^3 + x^2 + x + 1)f(x) + 0.
\end{aligned}
$$

Thus in the quotient ring $R = \frac{\mathbb{Z}_2[x]}{\langle f(x) \rangle}$, the product of these two nonzero elements is 0, and R is not an integral domain. $\qquad\Box$

THEOREM 2.27 The quotient ring $\frac{\mathbb{F}[x]}{\langle f(x) \rangle}$ is a field if and only if $f(x)$ is irreducible in $\mathbb{F}[x]$.

The *Lagrange Interpolation Formula* is a well-known method for constructing a polynomial based on given values for evaluation of a function.

THEOREM 2.28 *Lagrange Interpolation Formula.* Given $n + 1$ pairs $(a_i, b_i) \in \mathbb{F} \times \mathbb{F}$, with $a_i \neq a_j$, there exists a unique polynomial $f(x) \in \mathbb{F}[x]$ of degree at most n with $f(a_i) = b_i$. This polynomial is given by

$$f(x) = \sum_{i=0}^{n} b_i \prod_{\substack{k=0 \\ k \neq i}}^{n} \left(\frac{x - a_k}{a_i - a_k} \right).$$

Multivariate polynomial rings

Let $\mathbb{N}^n = \{(\alpha_1, \ldots, \alpha_n) \mid \alpha_i \in \mathbb{N}\}$ denote the set of *multi-indices* of size n. A *monomial* in the variables $x_1, \ldots, x_n$ is a product of the form

$$x_1^{\alpha_1} x_2^{\alpha_2} \ldots x_n^{\alpha_n},$$

which we denote by X^α, $\alpha \in \mathbb{N}^n$. The *degree* of X^α is $d_\alpha = \sum_{i=1}^{n} \alpha_i$.

DEFINITION 2.29 A *polynomial* in n variables $x_1, \ldots, x_n$ over the field $\mathbb{F}$ is a finite linear combination over $\mathbb{F}$ of monomials in $x_1, \ldots, x_n$, that is a formal expression of the form

$$\sum_{\alpha \in N} c_\alpha X^\alpha,$$

where $c_\alpha \in \mathbb{F}$ and N is a finite subset of $\mathbb{N}^n$.

DEFINITION 2.30 The set of all polynomials in n variables over a field $\mathbb{F}$ forms a ring under the standard operations of polynomial addition and multiplication. This ring is called a *polynomial ring* over $\mathbb{F}$, and for variables $x_1, \ldots, x_n$ is denoted by $\mathbb{F}[x_1, \ldots, x_n]$.

Let $f = \sum c_\alpha X^\alpha \in \mathbb{F}[x_1, \ldots, x_n]$ be a multivariate polynomial over $\mathbb{F}$. The summands $c_\alpha X^\alpha$ ($c_\alpha \neq 0$) are called the *terms* of f, and c_α is said to be the *coefficient* of X^α. The *total degree* of f is the maximum of the degrees of all monomials of f. If all monomials of f have the same degree d, we say that f is *homogeneous* of degree d.

DEFINITION 2.31 Let $f \in \mathbb{F}[x_1, \ldots, x_n]$ be a polynomial of total degree d. The polynomial f^h defined as

$$f^h = x_0^d \cdot f \left(\frac{x_1}{x_0}, \ldots, \frac{x_n}{x_0} \right)$$

is a homogeneous polynomial of degree d in $\mathbb{F}[x_0, x_1, \ldots, x_n]$, called the *homogenisation* of f.

DEFINITION 2.32 A total ordering $\prec$ on the set of monomials X^α (where $\alpha \in \mathbb{N}^n$) that is compatible with multiplication is called a *monomial ordering* in $\mathbb{F}[x_1, \ldots, x_n]$. An ordering is compatible with multiplication if $X^\alpha \prec X^\beta$ implies $X^\alpha X^\gamma \prec X^\beta X^\gamma$ for all multi-indices $\alpha, \beta, \gamma \in \mathbb{N}^n$.

We now define three common examples of monomial orderings.

DEFINITION 2.33 The *lex* (lexicographic) monomial ordering is defined by $X^\alpha \prec X^\beta$ if the left-most nonzero entry in the vector $\beta - \alpha \in \mathbb{Z}^n$ is positive.

DEFINITION 2.34 The *glex* (graded lexicographic) monomial ordering is defined by $X^\alpha \prec X^\beta$ if, firstly the degree of X^β is larger than the degree of X^α ($d_\beta > d_\alpha$), and secondly if $d_\beta = d_\alpha$ then the left-most nonzero entry in the vector $\beta - \alpha \in \mathbb{Z}^n$ is positive.

DEFINITION 2.35 The *grevlex* (graded reverse lexicographic) monomial ordering is defined by $X^\alpha \prec X^\beta$ if, firstly the degree of X^β is larger than the degree of X^α ($d_\beta > d_\alpha$), and secondly if $d_\beta = d_\alpha$ then the right–most nonzero entry in the vector $\beta - \alpha \in \mathbb{Z}^n$ is negative.

EXAMPLE 2.36 Some monomial orderings in $\mathbb{F}[x, y, z]$ are shown below.

$$\begin{aligned} \textit{lex ordering:} &\quad x^2 y^3 z^6 \prec x^2 y^4 z \quad \text{and} \quad xy^3 z \prec x^2 yz^2 \\ \textit{glex ordering:} &\quad x^2 y^4 z \prec x^2 y^3 z^6 \quad \text{and} \quad xy^3 z \prec x^2 yz^2 \\ \textit{grevlex ordering:} &\quad x^2 y^4 z \prec x^2 y^3 z^6 \quad \text{and} \quad x^2 yz^2 \prec xy^3 z \end{aligned}$$

We can see that the pair of monomials $x^2 y^4 z$ and $x^2 y^3 z^6$ and the pair of monomials $x^2 yz^2$ and $xy^3 z$ are ordered differently under the various monomial orderings. □

Suppose the polynomial ring $\mathbb{F}[x_1, \ldots, x_n]$ has a monomial ordering $\prec$ and $f \in \mathbb{F}[x_1, \ldots, x_n]$ is a polynomial. The *leading monomial* of f is the maximal monomial of f with respect to the ordering $\prec$ and is denoted by $\mathrm{LM}(f)$. The *leading coefficient* of f is the coefficient of the leading monomial of f and is denoted by $\mathrm{LC}(f)$. The *leading term* of f is the term associated with the leading monomial and is denoted by $\mathrm{LT}(f)$, so $\mathrm{LT}(f) = LC(f)LM(f)$. The *multidegree* of f is the degree of the leading monomial of f and is denoted by $\mathrm{multideg}(f)$.

These concepts enable us to give a multivariate generalisation of the division algorithm for univariate polynomials (Theorem 2.23).

THEOREM 2.37 *Polynomial Division Algorithm.* Suppose that the polynomial ring $\mathbb{F}[x_1, \ldots, x_n]$ has a monomial ordering $\prec$ and that $(g_1, \ldots, g_s)$ is an ordered subset of $\mathbb{F}[x_1, \ldots, x_n]$. For any $f \in \mathbb{F}[x_1, \ldots, x_n]$, there exist $a_i, r \in \mathbb{F}[x_1, \ldots, x_n]$ such that

$$f = a_1 g_1 + \ldots + a_s g_s + r,$$

where either $r = 0$, or $r \neq 0$ and no leading monomial of the polynomials g_i divides any of the monomials of r. Such a polynomial r is called a *remainder* of the division of f by the set $\{g_1, \ldots, g_s\}$. Furthermore, if $a_i g_i \neq 0$, then $\mathrm{multideg}(a_i g_i) \leq \mathrm{multideg}(f)$.

3. Linear Algebra

Linear algebra is at the heart of both the design and the analysis of the AES. Diffusion in the AES SP-network is achieved by a linear transformation. It is therefore not surprising to find linear algebra being used as a tool in the analysis of the cipher.

Vector spaces

DEFINITION 2.38 Let $(V, +)$ be an abelian group, $\mathbb{F}$ a field and $\cdot$ an operation $\mathbb{F} \times V \to V$. We say that V is a *vector space* over $\mathbb{F}$ if the following conditions hold.

- $a \cdot (v + v') = a \cdot v + a \cdot v'$ for all $v, v' \in V$ and $a \in \mathbb{F}$.

- $(a + a') \cdot v = a \cdot v + a' \cdot v$ for all $v \in V$ and $a, a' \in \mathbb{F}$.

- $(aa') \cdot v = a \cdot (a' \cdot v)$ for all $v \in V$ and $a, a' \in \mathbb{F}$.

- $1 \cdot v = v$ for all $v \in V$, where 1 is the identity element of $\mathbb{F}$.

In a vector space, an element of the set V is called a *vector* and an element of the field $\mathbb{F}$ is called a *scalar*. The operation $+$ is known as *vector addition* and the operation $\cdot$ as *scalar multiplication*. The identity element of the abelian group $(V, +)$ is called the *zero vector* and is usually denoted by 0. Furthermore, the symbol $\cdot$ is usually omitted if there is no danger of confusion.

EXAMPLE 2.39 The set $\mathbb{F}^n = \{(a_1, \ldots, a_n) \mid a_i \in \mathbb{F}\}$ forms a vector space over $\mathbb{F}$ with vector addition and scalar multiplication defined by

$$
\begin{aligned}
(a_1, \ldots, a_n) + (a'_1, \ldots, a'_n) &= (a_1 + a'_1, \ldots, a_n + a'_n), \text{ and} \\
a \cdot (a_1, \ldots, a_n) &= (aa_1, \ldots, aa_n).
\end{aligned}
$$
□

A subset U of a vector space V over a field $\mathbb{F}$ is a *subspace* of V if U is itself a vector space over $\mathbb{F}$. The notation $U < V$ is used to denote that U is a subspace of V. The intersection $U \cap U'$ of any two subspaces U and U' of V is a subspace of V. The sum of subspaces $U, U' < V$, defined by

$$ U + U' = \{u + u' \mid u \in U, u' \in U'\}, $$

is also a subspace of V. This definition extends in the obvious way to any finite sum of subspaces. If a vector space $V = U_1 + \ldots + U_m$ and the subspaces $U_1, \ldots, U_m$ have trivial pairwise intersections ($U_i \cap U_j = \{0\}$ for $i \neq j$), then V is said to be the *direct sum* of these subspaces and we write $V = U_1 \oplus \ldots \oplus U_m$. In this case, for any $v \in V$, there exist unique $u_i \in U_i$ such that $v = u_1 + \ldots + u_m$.

The set of all finite linear combinations of the vectors $v_1, \ldots, v_m \in V$,

$$ \langle v_1, \ldots, v_m \rangle = \{a_1 v_1 + \ldots + a_m v_m \mid a_i \in \mathbb{F}, v_i \in V\}, $$

is a subspace of V and is called the *subspace generated* by the set $\{v_1, \ldots, v_m\}$. A set of vectors $\{v_1, \ldots, v_m\}$ is said to *span* or to be a *spanning set* of a subspace $U < V$ if for all $u \in U$ there exists $a_i \in \mathbb{F}$ such that $u = \sum_i a_i v_i$.

A *basis* for a vector space V is a minimal spanning set for V. Every vector in a vector space can be expressed as a unique linear combination over $\mathbb{F}$ of the elements of a basis. Any basis for a vector space V always has the same cardinality, which is called the *dimension* of the vector space V and is denoted by $\dim(V)$.

EXAMPLE 2.40 The multivariate polynomial ring $\mathbb{F}[x_1, \ldots, x_n]$ forms an infinite dimensional vector space over $\mathbb{F}$. The subset of $\mathbb{F}[x_1, \ldots, x_n]$ of all polynomials of degree at most 1 is a subspace of dimension $n+1$ with basis $\{1, x_1, \ldots, x_n\}$. □

A set of vectors $\{v_1, \ldots, v_m\}$ is said to be *linearly independent* if the expression $\sum_{i=1}^{m} a_i v_i = 0$ implies that $a_1 = \ldots = a_m = 0$. If $\{e_1, \ldots, e_n\}$ is a basis B for a vector space V of dimension n, then B is linearly independent and for any $v \in V$ there exist unique $a_i \in \mathbb{F}$ such that $v = a_1 e_1 + \ldots + a_n e_n$. Thus we can represent v with respect to the basis B by the n-tuple $(a_1, \ldots, a_n) \in \mathbb{F}^n$.

We can define cosets of subspaces in a similar manner to cosets of subgroups. In particular, the set of all cosets of U in V forms a vector space called a *quotient vector space* and is denoted by V/U.

Linear transformations

DEFINITION 2.41 A *linear transformation* or vector space *homomorphism* from a vector space V over a field $\mathbb{F}$ to a vector space U over $\mathbb{F}$ is a mapping $\psi: V \to U$ that satisfies the following two conditions.

- $\psi(v + v') = \psi(v) + \psi(v')$ for all $v, v' \in V$.

- $\psi(av) = a \cdot \psi(v) = a\psi(v)$ for all $v \in V$ and $a \in \mathbb{F}$.

A vector space *isomorphism* is a bijective linear transformation, and we use $V \cong U$ to denote that the vector spaces V and U are isomorphic.

EXAMPLE 2.42 Let V be a vector space over the field $\mathbb{F}$ of dimension n, and let $\{e_1, \ldots, e_n\}$ be a basis for V. Given $v \in V$, there exists a unique $(a_1, \ldots, a_n) \in \mathbb{F}^n$ such that $v = a_1 e_1 + \ldots + a_n e_n$. The mapping $V \to \mathbb{F}^n$ defined by $v \mapsto (a_1, \ldots, a_n)$ is a vector space isomorphism. Thus any two finite–dimensional vector spaces over the same field are isomorphic. □

EXAMPLE 2.43 Let V be a vector space of dimension n over the field $\mathbb{F}$, and $\alpha_1, \ldots, \alpha_n$ be elements of $\mathbb{F}$. Then every mapping $V \to \mathbb{F}$ of the form $a_1 e_1 + \ldots + a_n e_n \mapsto \alpha_1 a_1 + \ldots + \alpha_n a_n$ is a linear transformation from V into $\mathbb{F}$, where $\mathbb{F}$ is considered as a one–dimensional vector space over $\mathbb{F}$. Furthermore, every linear transformation from $V \to \mathbb{F}$ is of this form. Such a transformation is known as a *linear functional* on V. □

Let $\psi: V \to V'$ be a linear transformation. Then the *kernel* of ψ is defined by $\ker \psi = \{v \in V \,|\, \psi(v) = 0\}$ and is a subspace of V. The *nullity* of ψ is the dimension of $\ker \psi$. The image of the linear transformation ψ is a subspace of V', and the *rank* of ψ is the dimension of $\psi(V)$. The *Rank–Nullity Theorem* states that

$$\dim(V) = \dim(\ker \psi) + \dim(\psi(V)).$$

The quotient vector space $V/\ker \psi$ is isomorphic to the image of V, so $\psi(V) \cong V/\ker \psi$. If $V' = V$, then the subspace $U < V$ is called a *ψ-invariant* subspace if $\psi(U) < U$. If ψ satisfies $\psi \circ \psi = \psi^2 = \psi$, then ψ is called a *projection* and $\psi(U)$ is a ψ-invariant subspace for any subspace $U < V$.

If $\psi: V \to V$ is a linear transformation, then $\sum_{i=0}^{d} a_i \psi^i$ is also a linear transformation on V. Furthermore, the set

$$I = \left\{ \sum_{i=0}^{d} a_i x^i \ \middle|\ \sum_{i=0}^{d} a_i \psi^i = 0 \right\}$$

is an ideal of the polynomial ring $\mathbb{F}[x]$. The *minimal polynomial* of the linear transformation ψ is defined as the unique monic polynomial $\min_\psi(x)$ that generates the principal ideal I. The minimal polynomial of ψ gives much information about both ψ and the ψ-invariant subspaces. For example, if $\min_\psi(x) = m_1(x) \ldots m_l(x)$ is the factorisation of the minimal polynomial of ψ into monic polynomials, then $m_i(\psi)$ has a natural interpretation as a linear transformation and the ψ-invariant subspaces are given by $\ker m_i(\psi)$.

DEFINITION 2.44 Suppose that V and V' are vector spaces over the field $\mathbb{F}$ and that $\psi: V \to V'$ is a linear transformation. If b is a vector in V', then the transformation $V \to V'$ defined by $v \mapsto \psi(v) + b$ is termed an *affine transformation*.

DEFINITION 2.45 Consider a mapping $\beta : V \times V \to V'$, where V and V' are vector spaces over the field $\mathbb{F}$. For $u \in V$, we can define the mappings $\beta'_u, \beta''_u: V \to V'$ by $v \mapsto \beta(u, v)$ and $v \mapsto \beta(v, u)$ respectively. The mapping β is called a *bilinear transformation* on V if β'_u and β''_u are linear transformations for all $u \in V$.

Matrices

DEFINITION 2.46 An $m \times n$ *matrix* over a field $\mathbb{F}$ is a rectangular array

$$\begin{pmatrix} a_{11} & \cdots & a_{1n} \\ \vdots & \ddots & \vdots \\ a_{m1} & \cdots & a_{mn} \end{pmatrix}$$

with $a_{ij} \in \mathbb{F}$. The elements a_{ij} are called the *entries* of the matrix.

If A is an $m \times n$ matrix, the sequences $(a_{i1} \ldots a_{in})$ are called the *rows* of A, and the sequences $(a_{1j} \ldots a_{mj})$ are called the *columns* of A. Thus A has m rows and n columns. If $m = n$, then A is called a *square matrix* of order n. A *submatrix* of A is an $m' \times n'$ matrix ($m' \le m$ and $n' \le n$) obtained by taking a block of entries of M with m' rows and n' columns. The *transpose* of A is denoted by A^T and is the $n \times m$ matrix whose (i, j)-entry is given by a_{ji}.

EXAMPLE 2.47 Let $\mathcal{M}_{m \times n}(\mathbb{F})$ denote the set of all $m \times n$ matrices over $\mathbb{F}$. We can define the operation of addition of elements of $\mathcal{M}_{m \times n}(\mathbb{F})$ in the obvious way by adding the corresponding entries of the matrices. Similarly, we can define the scalar multiplication of a matrix $A \in \mathcal{M}_{m \times n}(\mathbb{F})$ by an element $c \in \mathbb{F}$ to be the matrix obtained by simply multiplying every entry of A by c. Thus the set $\mathcal{M}_{m \times n}(\mathbb{F})$ forms a vector space over $\mathbb{F}$ of dimension mn. □

Let A be an $m \times n$ matrix and B be an $r \times s$ matrix over $\mathbb{F}$ defined as

$$A = \begin{pmatrix} a_{11} & \cdots & a_{1n} \\ \vdots & \ddots & \vdots \\ a_{m1} & \cdots & a_{mn} \end{pmatrix}, \quad B = \begin{pmatrix} b_{11} & \cdots & b_{1s} \\ \vdots & \ddots & \vdots \\ b_{r1} & \cdots & b_{rs} \end{pmatrix}.$$

If $n = r$, we can define a multiplication of A by B. The *product* AB is the $m \times s$ matrix C whose entries are $c_{ij} = \sum_{k=1}^{n} a_{ik} b_{kj}$.

DEFINITION 2.48 Suppose A is an $m \times n$ matrix over $\mathbb{F}$ and that $A_{i\cdot}$ denotes the ith row of A. An *elementary row operation* on the matrix A is one of the following three types of operation.

- The replacement of $A_{i\cdot}$ by $cA_{i\cdot}$ where $c \in \mathbb{F}$ with $c \ne 0$.

- The replacement of $A_{i\cdot}$ by $A_{i\cdot} + cA_{j\cdot}$ where $c \in \mathbb{F}$ and $i \ne j$.

- The interchange of two rows of A.

An elementary row operation on the matrix A is equivalent to a mapping $A \mapsto PA$, where P is a $m \times m$ elementary row operation matrix.

Any $m \times n$ matrix that can be obtained from A by a series of elementary row operations is said to be *row–equivalent* to A. In particular, there is a special set of matrices called the *row–reduced echelon matrices*, and any matrix is row–equivalent to a unique row–reduced echelon matrix.

The *rank* of an $m \times n$ matrix A is the number of linearly independent rows or columns (considered as vectors) of A. In particular, if A is a row–reduced echelon matrix, then the rank of A is the number of nonzero rows. We note that row–equivalent matrices have the same rank.

Let $\mathcal{M}_n(\mathbb{F})$ denote the set of all square matrices over $\mathbb{F}$ of order n. A matrix $A \in \mathcal{M}_n(\mathbb{F})$ with entries a_{ij} is a *symmetric matrix* if $A^T = A$, that is $a_{ij} = a_{ji}$, and A is a *diagonal matrix* if $a_{ij} = 0$ whenever $i \neq j$. The *identity matrix* is a diagonal matrix in which $a_{ii} = 1$ $(i = 1, \ldots, n)$ and is usually denoted by I. The identity matrix has the property that $AI = IA = A$ for any matrix $A \in \mathcal{M}_n(\mathbb{F})$. The square matrix A is an *invertible* or *non–singular* matrix if there exists an $n \times n$ matrix A^{-1} such that $AA^{-1} = A^{-1}A = I$. A matrix is invertible if and only if it is row–equivalent to the identity matrix.

The *determinant* is a function $\det \colon \mathcal{M}_n(\mathbb{F}) \to \mathbb{F}$ on square matrices with special properties, and this function is widely used in the analysis of square matrices [59]. In particular, we have $\det(AB) = \det(A)\det(B)$, and a matrix A is invertible if and only if $\det(A) \neq 0$.

The set of $n \times n$ invertible matrices forms a group under the operation of matrix multiplication. This group is called the *general linear group* and is denoted by $\mathrm{GL}(n, \mathbb{F})$. The subset of all matrices that have determinant 1 forms a normal subgroup of $\mathrm{GL}(n, \mathbb{F})$. This subgroup is called the *special linear group* and is denoted by $\mathrm{SL}(n, \mathbb{F})$. Thus we have

$$
\begin{aligned}
\mathrm{GL}(n, \mathbb{F}) &= \{\, A \in \mathcal{M}_n(\mathbb{F}) \mid \det(A) \neq 0 \,\}, \text{ and} \\
\mathrm{SL}(n, \mathbb{F}) &= \{\, A \in \mathcal{M}_n(\mathbb{F}) \mid \det(A) = 1 \,\}.
\end{aligned}
$$

Matrices are often used to represent linear transformations between vector spaces and can be particularly useful for performing calculations with such mappings. For example, matrices provide an easy way of calculating the image of vectors under linear transformations or of calculating the composition of linear transformations. Furthermore, many properties of a linear transformation, such as its rank, minimal polynomial, invariant subspaces, can be easily obtained by analysing a matrix corresponding to that linear transformation.

Suppose that $\psi : V \to V'$ is a linear transformation between two vector spaces V and V' over a field $\mathbb{F}$ of dimensions n and m respectively. Suppose further that V has a basis $B = \{e_1, \ldots, e_n\}$ and V' has a basis $B' = \{e'_1, \ldots, e'_m\}$. Then there exist $a_{ij} \in \mathbb{F}$ such that $\psi(e_i) = \sum_{j=1}^{m} a_{ij} e'_j$ $(1 \leq i \leq n)$, and the matrix of the linear transformation ψ

with respect to the bases B and B' is defined as the $m \times n$ matrix A whose entries are a_{ij}. Any $v \in V$ is given by $v = \sum_{j=1}^{n} v_j e_j$ for some $v_i \in \mathbb{F}$. In this monograph, we represent vectors as *column vectors* or $n \times 1$ matrices. Thus the vector v is given by the column vector $\begin{pmatrix} v_1 \\ \vdots \\ v_n \end{pmatrix}$ with respect to the basis B, which we write as $(v_1, \ldots, v_n)^T$. The effect of the linear transformation ψ on the vector v is given by

$$\psi(v) = \sum_{j=1}^{n} v_j \psi(e_j) = \sum_{j=1}^{n} v_i \left(\sum_{i=1}^{m} a_{ij} e_i' \right) = \sum_{i=1}^{m} \left(\sum_{j=1}^{n} a_{ij} v_j \right) e_i',$$

which is expressed in terms of matrices by the matrix multiplication

$$Av = \begin{pmatrix} a_{11} & \cdots & a_{1n} \\ \vdots & \ddots & \vdots \\ a_{m1} & \cdots & a_{mn} \end{pmatrix} \begin{pmatrix} v_1 \\ \vdots \\ v_n \end{pmatrix} = \begin{pmatrix} a_{11}v_1 + \ldots + a_{1n}v_n \\ \vdots \\ a_{m1}v_1 + \ldots + a_{mn}v_n \end{pmatrix}.$$

The composition of linear transformations can also be easily computed using matrices. If $\psi : V \to V'$ and $\psi' : V' \to V''$ are linear transformations, and A and A' are the matrices associated with ψ and ψ' respectively, then $P = A'A$ is the matrix associated with the linear transformation $\psi' \circ \psi : V \to V''$. Thus the matrix of a composition of linear transformations is the product of the respective matrices.

We note that the matrix corresponding to a linear transformation is not unique as it depends on the basis chosen for the vector spaces. Suppose, as above, that we have a linear transformation $\psi : V \to V'$ between two vector spaces V and V' of dimension m and n respectively. If the linear transformation ψ is represented by an $m \times n$ matrix A with respect to one pair of bases and by another $m \times n$ matrix $\widetilde{A}$ with respect to another pair of bases, then there exist an invertible $n \times n$ matrix P and an invertible $m \times m$ matrix P' such that $\widetilde{A} = P'AP$. We say that the matrix $\widetilde{A}$ is obtained from A by a *change of basis*.

DEFINITION 2.49 Let A be an $n \times n$ matrix over the field $\mathbb{F}$. The *minimal polynomial* of the matrix A is the unique monic polynomial $\min_A(x) \in \mathbb{F}[x]$ of minimal degree such that $\min_A(A) = 0$. The *characteristic polynomial* of A is the polynomial $c_A(x) \in \mathbb{F}[x]$ defined by

$$c_A(x) = \det(xI - A).$$

We note that the minimal polynomial of a linear transformation is the same as the minimal polynomial of any of its associated matrices.

THEOREM 2.50 *Cayley–Hamilton Theorem.* The minimal polynomial of a matrix divides the characteristic polynomial.

EXAMPLE 2.51 Consider the matrix

$$A = \begin{pmatrix} 0 & 1 & 0 & 0 \\ 1 & 0 & 0 & 1 \\ 0 & 0 & 1 & 1 \\ 1 & 1 & 0 & 0 \end{pmatrix}$$

over the field $\mathbb{Z}_2$. It can be shown that the minimal polynomial of A is $\min_A(x) = x^3 + 1$ and the characteristic polynomial of A is

$$c_A(x) = x^4 + x^3 + x + 1.$$

We note that the minimal polynomial $\min_A(x)$ divides $c_A(x)$. Furthermore, $\min_A(x) = (x + 1)(x^2 + x + 1)$ as a product of irreducible polynomials. Thus, if $\psi : V \to V'$ is a linear transformation associated with A, then the invariant subspaces of ψ are given by $\ker(\psi_1)$ and $\ker(\psi_2)$, where ψ_1 and ψ_2 are the linear transformations $V \to V'$ associated with the matrices $(A + I)$ and $(A^2 + A + I)$ respectively.   □

Matrices are also widely used in coding theory, and most properties of linear codes can obtained by studying their generator and parity check matrices. Of special interest in the design and analysis of the AES are the matrices that arise from *maximal distance separable (MDS) codes* [76].

DEFINITION 2.52 An $m \times n$ matrix A is called an MDS matrix if and only if every square submatrix of A is invertible.

Linear systems and matrix complexity

Matrices can be used to represent systems of linear equations. Suppose we have such a system of m equations in n variables $x_1, \ldots, x_n$ given by

$$a_{11}x_1 + \ldots + a_{1n}x_n = b_1$$
$$\vdots \qquad\qquad \vdots$$
$$a_{m1}x_1 + \ldots + a_{mn}x_n = b_m,$$

where a_{ij} and b_i are elements of a field $\mathbb{F}$. This equation system can be represented by the matrix equation

$$\begin{pmatrix} a_{11} & \cdots & a_{1n} \\ \vdots & \ddots & \vdots \\ a_{m1} & \cdots & a_{mn} \end{pmatrix} \begin{pmatrix} x_1 \\ \vdots \\ x_n \end{pmatrix} = \begin{pmatrix} b_1 \\ \vdots \\ b_m \end{pmatrix},$$

or equivalently $Ax = b$. The standard process of solving this equation system is to transform the matrix A to a row–reduced echelon matrix using elementary row operations. This corresponds to finding an invertible $m \times m$ matrix P such that PA is a row–reduced echelon matrix. This allows us to obtain an equivalent matrix equation $PAx = Pb$, which gives us an immediate full solution for $x_1, \ldots, x_n$.

The simplest method of transforming the matrix A to a row–reduced echelon matrix is known as *Gaussian reduction*. Performing Gaussian reduction on a square $n \times n$ matrix takes of the order of n^3 field operations. However, more sophisticated techniques for row–reducing a matrix can reduce this to less than cubic complexity.

DEFINITION 2.53 An $n \times n$ square matrix can be transformed to a row–reduced echelon matrix with complexity of the order of n^ω field operations. We call ω the *exponent of matrix reduction*. Thus $\omega = 3$ for Gaussian reduction. The smallest values of ω occur for row–reduction techniques for a *sparse matrix*, that is a matrix whose almost all entries are zero. The exponent of matrix reduction ω satisfies $2 < \omega \leq 3$.

Algebras

DEFINITION 2.54 Suppose $\mathcal{A}$ is a vector space over a field $\mathbb{F}$ with a multiplication operation $\mathcal{A} \times \mathcal{A} \to \mathcal{A}$. If this multiplication operation is associative and a bilinear mapping on the vector space $\mathcal{A}$, then $\mathcal{A}$ is an (associative) $\mathbb{F}$-*algebra*, or more simply an *algebra*.

Informally, we can regard an algebra as a vector space that is also a ring. The dimension of the algebra $\mathcal{A}$ is the dimension of $\mathcal{A}$ as a vector space. The subset $\mathcal{A}' \subset \mathcal{A}$ is a *subalgebra* of $\mathcal{A}$ if $\mathcal{A}'$ is an algebra in its own right, and $\mathcal{A}'$ is an *ideal subalgebra* if it is also an ideal of the ring $\mathcal{A}$. We can also classify mappings between two algebras in the usual way, so an *algebra homomorphism* is a mapping that is both a ring homomorphism and a vector space homomorphism.

EXAMPLE 2.55 The ring of polynomials $\mathbb{F}[x_1, \ldots, x_n]$ is a vector space over $\mathbb{F}$ (Example 2.40). Thus $\mathbb{F}[x_1, \ldots, x_n]$ forms an $\mathbb{F}$-algebra, known as a *polynomial algebra*. □

EXAMPLE 2.56 The set $\mathcal{M}_n(\mathbb{F})$ of $n \times n$ matrices over $\mathbb{F}$ forms a vector space over $\mathbb{F}$ of dimension n^2 (Example 2.47). Matrix multiplication is an associative bilinear mapping on $\mathcal{M}_n(\mathbb{F})$. Thus $\mathcal{M}_n(\mathbb{F})$ forms an $\mathbb{F}$-algebra of dimension n^2. The set $\mathcal{D}_n(\mathbb{F})$ of $n \times n$ diagonal matrices over $\mathbb{F}$ forms a subalgebra of $\mathcal{M}_n(\mathbb{F})$ of dimension n. Such algebras are known as *matrix algebras*. □

4.   Finite Fields

The design of the AES is based around finite fields. All the operations used by the AES are described by algebraic operations on a finite field of even characteristic. In this section, we discuss the properties of finite fields relevant to the specification and algebraic analysis of the AES.

Finite fields and subfields

The set $\mathbb{Z}_p = \{0, \ldots, p-1\}$ with addition and multiplication operations defined modulo p forms a finite field if and only if p is prime (Example 2.20). This field is called the *Galois field* of order p and is denoted by $\mathrm{GF}(p)$. The Galois field $\mathrm{GF}(p)$ plays a fundamental role in the theory of finite fields.

DEFINITION 2.57 Suppose that $\mathbb{F}$ and $\mathbb{K}$ are two fields. If $\mathbb{F} \subset \mathbb{K}$, then $\mathbb{F}$ is said to be a *subfield* of $\mathbb{K}$, or equivalently $\mathbb{K}$ is said to be an *extension field* of $\mathbb{F}$.

THEOREM 2.58 A finite field of characteristic p (prime) has a unique minimal subfield isomorphic to $\mathrm{GF}(p)$.

If $\mathbb{K}$ is a extension field of the field $\mathbb{F}$, then $\mathbb{K}$ is also a vector space over $\mathbb{F}$. The dimension of this vector space is the *degree* of the extension. If $\mathbb{F}$ has order q and $\mathbb{K}$ is an extension field of $\mathbb{F}$ of degree d, then $\mathbb{K}$ has order q^d. As every finite field has prime characteristic p, it follows from Theorem 2.58 that every finite field has order p^n for some prime p and some positive integer n.

THEOREM 2.59 For every prime number p and every positive integer n, there exists a finite field of order p^n. Furthermore, any two finite fields of order p^n are isomorphic.

Thus finite fields of order p^n are unique up to isomorphism. This field is called the *Galois field* of order p^n and denoted by $\mathrm{GF}(p^n)$. A subfield of $\mathrm{GF}(p^n)$ has order p^d, where d is a divisor of n. Furthermore, there is exactly one subfield of order p^d for every divisor d of n. For example, the finite field $\mathrm{GF}(2^8)$ has $\mathrm{GF}(2^4)$, $\mathrm{GF}(2^2)$, and $\mathrm{GF}(2)$ as proper subfields.

THEOREM 2.60 The multiplicative group $\mathrm{GF}(q)^*$ is a cyclic group of order $q - 1$.

A generator of the multiplicative group $\mathrm{GF}(q)^*$ is called a *primitive element* of the field $\mathrm{GF}(q)$. The number of primitive elements in $\mathrm{GF}(q)$ is $\varphi(q - 1)$, where $\varphi(m)$ is *Euler's totient function*, which gives the number of positive integers less than or equal to m and coprime to m.

Explicit construction of finite fields

Theorem 2.27 provides a method of constructing a finite field as a quotient ring. Suppose $\mathbb{F}$ is a finite field of order $q = p^n$ and $f(x) \in \mathbb{F}[x]$ is an irreducible polynomial of degree d. The quotient ring $\mathbb{K} = \frac{\mathbb{F}[x]}{\langle f(x) \rangle}$ is a field of order $q^d = p^{nd}$, which is an extension field of degree d of $\mathbb{F}$. In the manner given in Example 2.25, its elements can be represented as

$$a_{d-1} x^{d-1} + \ldots + a_2 x^2 + a_1 x + a_0,$$

where $a_i \in \mathbb{F}$. Addition and multiplication are then as described in Example 2.25. Theorem 2.59 states that any finite field of order p^{nd} is isomorphic to $\mathbb{K}$.

We can also construct $\mathrm{GF}(p^{nd})$ directly as an *extension field* of $\mathbb{F}$. Let θ denote a root of the irreducible polynomial $f(x)$ of degree d. The set $\mathbb{F}(\theta)$ of all quotients (with nonzero denominator) of polynomials in θ with coefficients in $\mathbb{F}$ is the smallest field containing both θ and $\mathbb{F}$. Furthermore, $\mathbb{F}(\theta)$ is the extension field obtained by *adjoining* θ to $\mathbb{F}$. This extension field $\mathbb{F}(\theta)$ has p^{nd} elements and so is isomorphic to $\mathrm{GF}(p^{nd})$. The elements of $\mathbb{F}(\theta)$ are given by

$$a_{d-1} \theta^{d-1} + \ldots + a_2 \theta^2 + a_1 \theta + a_0,$$

where $a_i \in \mathbb{F}$. If the element θ is a generator of the multiplicative group of $\mathbb{F}(\theta)$, then the polynomial $f(x)$ is called a *primitive polynomial*.

EXAMPLE 2.61 The polynomial $m(x) = x^8 + x^4 + x^3 + x + 1 \in \mathrm{GF}(2)[x]$ is irreducible. If θ is a root of $m(x)$, then

$$\mathrm{GF}(2)(\theta) \cong \frac{\mathrm{GF}(2)[x]}{\langle m(x) \rangle} \cong \mathrm{GF}(2^8).$$

The elements of the quotient ring $\frac{\mathrm{GF}(2)[x]}{\langle m(x) \rangle}$ are given, for $a_i \in \mathbb{F}$, by

$$a_7 x^7 + \ldots + a_2 x^2 + a_1 x + a_0;$$

whereas the elements of extension field $\mathbb{F}(\theta)$ are given, for $b_i \in \mathbb{F}$, by

$$b_7 \theta^7 + \ldots + b_2 \theta^2 + b_1 \theta + b_0.$$

We note that $m(x)$ is not primitive, since the order of $\theta \in \mathbb{F}(\theta)$ is 51. $\square$

Irreducible polynomials over a field $\mathbb{F}$ of order q are the basic tools for the construction of all finite extensions of $\mathbb{F}$. If $\mathbb{K}$ is an extension of $\mathbb{F}$ of order q^n, then Theorem 2.60 shows that $a^{q^n - 1} - 1 = 0$ for all nonzero $a \in \mathbb{K}$. Thus the polynomial $x^{q^n} - x$ has all q^n elements of $\mathbb{K}$ as a root. The field $\mathbb{K} \cong \mathrm{GF}(q^n)$ is known as the *splitting field* of the polynomial $x^{q^n} - x$. This polynomial can be used to obtain all irreducible polynomials over $\mathbb{F}$ with the required degree.

THEOREM 2.62 Let $\mathbb{F}$ be a finite field of order q. Then the polynomial $x^{q^n} - x \in \mathbb{F}[x]$ is the product of all monic irreducible polynomials in $\mathbb{F}[x]$ whose degree divides n.

The number of irreducible polynomials in $\mathbb{F}[x]$ of degree n is given by

$$\frac{1}{n} \sum_{d|n} \mu(d) q^{\frac{n}{d}},$$

where μ is the *Möbius function*, defined by $\mu(1) = 1$, $\mu(n) = (-1)^k$ if n is the product of k distinct primes, and 0 otherwise. The number of primitive polynomials of degree n is $\frac{1}{n}\varphi(q^n - 1)$, where φ is Euler's totient function.

EXAMPLE 2.63 There are $\frac{1}{8}\left(\mu(1)2^8 + \mu(2)2^4 + \mu(4)2^2 + \mu(8)2^1\right) = 60$ irreducible polynomials of degree 8 in $\mathrm{GF}(2)[x]$, of which $\frac{1}{8}\varphi(2^8 - 1) = 16$ are primitive polynomials.   □

DEFINITION 2.64 A field $\mathbb{F}$ is said to be *algebraically closed* if every polynomial in $\mathbb{F}[x]$ has a root in $\mathbb{F}$. The *algebraic closure* of a field $\mathbb{F}$ is the smallest extension field $\mathbb{K}$ of $\mathbb{F}$ such that $\mathbb{K}$ is algebraically closed.

Representations of a finite field

Let $\mathbb{F}$ be a field and $\mathbb{K} = \mathbb{F}(\theta)$ be an extension field of $\mathbb{F}$ of degree d. The most common way to describe the elements of $\mathbb{K}$ is to regard all elements as vectors in the vector space $\mathbb{K}$ of dimension d over the $\mathbb{F}$. Every element in $\mathbb{K}$ can be written uniquely as

$$a_{d-1}\theta^{d-1} + \ldots + a_2\theta^2 + a_1\theta + a_0,$$

where $a_i \in \mathbb{F}$. Thus the set $\{\theta^{d-1}, \ldots, \theta^2, \theta, 1\}$ forms a basis of $\mathbb{K}$ as a d-dimensional vector space over $\mathbb{F}$. This basis is called a *polynomial basis* for the field $\mathbb{K}$.

EXAMPLE 2.65 Suppose θ is a root of $x^8 + x^4 + x^3 + x + 1 \in \mathrm{GF}(2)[x]$, and let $\mathbb{K}$ be the field $\mathrm{GF}(2)(\theta)$ (Example 2.61). Any multiplication mapping $\mathbb{K} \to \mathbb{K}$ is a linear transformation of $\mathbb{K}$ as a vector space over $\mathrm{GF}(2)$. The squaring mapping in $\mathbb{K}$ is also a linear transformation. If we let T_θ and S denote the matrices that correspond to multiplication by θ and squaring with respect to the polynomial basis $\{\theta^7, \ldots, \theta^2, \theta, 1\}$, then we have

$$T_\theta = \begin{pmatrix} 0 & 1 & 0 & 0 & 0 & 0 & 0 & 0 \\ 0 & 0 & 1 & 0 & 0 & 0 & 0 & 0 \\ 0 & 0 & 0 & 1 & 0 & 0 & 0 & 0 \\ 1 & 0 & 0 & 0 & 1 & 0 & 0 & 0 \\ 1 & 0 & 0 & 0 & 0 & 1 & 0 & 0 \\ 0 & 0 & 0 & 0 & 0 & 0 & 1 & 0 \\ 1 & 0 & 0 & 0 & 0 & 0 & 0 & 1 \\ 1 & 0 & 0 & 0 & 0 & 0 & 0 & 0 \end{pmatrix} \quad S = \begin{pmatrix} 1 & 1 & 0 & 0 & 0 & 0 & 0 & 0 \\ 0 & 0 & 1 & 0 & 1 & 0 & 0 & 0 \\ 0 & 1 & 1 & 0 & 0 & 0 & 0 & 0 \\ 1 & 0 & 0 & 1 & 0 & 1 & 0 & 0 \\ 1 & 1 & 1 & 1 & 0 & 0 & 0 & 0 \\ 0 & 0 & 1 & 0 & 0 & 0 & 1 & 0 \\ 1 & 1 & 0 & 1 & 0 & 0 & 0 & 0 \\ 0 & 1 & 0 & 1 & 0 & 0 & 0 & 1 \end{pmatrix}.$$

$\square$

There are other bases which are used for the field $\mathbb{K}$ when considered as a vector space over $\mathbb{F}$, such as the *normal basis* $\{\beta, \beta^q, \beta^{q^2}, \ldots, \beta^{q^{d-1}}\}$ for suitable $\beta \in \mathbb{K}$. This representation is particularly useful when performing the exponentiation of elements in $\mathbb{K}$ and may offer implementation advantages in some situations.

There are also methods of describing an element of the finite field $\mathbb{K}$ of order q^n which depend on logarithmic functions of $\mathbb{K}$ rather than the vector space aspect of $\mathbb{K}$. Suppose β is a primitive element of $\mathbb{K}$ and $a = \beta^i$ $(0 \leq i < q^n - 1)$. The *discrete logarithm* is a function $\log_\beta : \mathbb{K}^* \to \mathbb{Z}_{q^n-1}$ defined by $\log_\beta a = \log_\beta \beta^i = i$. We can thus represent the nonzero elements $a \in \mathbb{K}$ by $\log_\beta a \in \mathbb{Z}_{q^n-1}$. If we adopt the convention that the discrete logarithm of 0 is denoted by ∞, then we can represent an element of $\mathbb{K}$ by an element of $\overline{\mathbb{Z}}_{q^n-1} = \mathbb{Z}_{q^n-1} \cup \{\infty\}$.

The *Zech* or *Jacobi logarithm* offers another logarithmic method for describing a finite field element. The Zech logarithm is based on the function $Z : \overline{\mathbb{Z}}_{q^n-1} \to \overline{\mathbb{Z}}_{q^n-1}$ given by

$$Z(n) = \log_\beta(\beta^n + 1),$$

so $\beta^{Z(n)} = \beta^n + 1$ with the convention that $\beta^\infty = 0$. The definition can be extended to all integers by working modulo $q^n - 1$. The Zech logarithm of β^n can now be defined to be $Z(n)$. We have the following identities concerning this function Z:

$$\begin{aligned} Z(Z(n)) &= n, \\ Z(2n) &= 2Z(n), \\ Z(-n) &= Z(n) - n. \end{aligned}$$

This function is of interest since it can be used to calculate the sum of two powers of β, since

$$\beta^m + \beta^n = \beta^n(\beta^{m-n} + 1) = \beta^n \beta^{Z(m-n)} = \beta^{n+Z(m-n)}.$$

Functions in a finite field

DEFINITION 2.66 Let $\mathbb{F}$ be a finite field of order q and $\mathbb{K}$ be an extension field of $\mathbb{F}$ of degree d. The elements $a, a^q, a^{q^2}, \ldots, a^{q^{d-1}}$ are the *conjugates* of $a \in \mathbb{K}$ with respect to $\mathbb{F}$.

THEOREM 2.67 Suppose $\mathbb{K}$ is an extension of a field $\mathbb{F}$ of degree d. Any element $a \in \mathbb{K}$ is a root of an irreducible polynomial $f(x) \in \mathbb{F}[x]$ of degree n dividing d. The roots of $f(x)$ are the conjugates of a.

We now consider some functions of interest on finite fields.

DEFINITION 2.68 Let $\mathbb{F}$ be a finite field of order q and $\mathbb{K}$ be an extension field of $\mathbb{F}$ of degree d. The *trace* function on $\mathbb{K}$ with respect to $\mathbb{F}$ is the function $\mathrm{Tr}\colon \mathbb{K} \to \mathbb{F}$ defined by

$$\mathrm{Tr}(a) = a + a^q + a^{q^2} + \ldots + a^{q^{d-1}}.$$

Thus the trace of an element $a \in \mathbb{K}$ is the sum of all conjugates of a. The trace function is a linear functional on $\mathbb{K}$, considered as a vector space over $\mathbb{F}$ (Example 2.43). In fact, any linear functional on $\mathbb{K}$ is of the form $a \mapsto \mathrm{Tr}(\beta a)$ for some $\beta \in \mathbb{K}$.

DEFINITION 2.69 Let $\mathbb{F}$ be a finite field of order q and $\mathbb{K}$ be an extension field of $\mathbb{F}$ of degree d. The *norm* function on $\mathbb{K}$ with respect to $\mathbb{F}$ is the function $\mathrm{N}\colon \mathbb{K} \to \mathbb{F}$ defined by

$$\mathrm{N}(a) = a \, a^q \, a^{q^2} \, \ldots \, a^{q^{d-1}} = a^{\frac{q^d - 1}{q - 1}}.$$

Thus the norm of an element $a \in \mathbb{K}$ is the product of all conjugates of a. The norm function is a group homomorphism $\mathbb{K}^* \to \mathbb{F}^*$ between the multiplicative groups of the fields $\mathbb{K}$ and $\mathbb{F}$.

DEFINITION 2.70 A *linearised polynomial* $f(x) \in \mathbb{K}[x]$ is a polynomial given by

$$f(x) = a_0 x + a_1 x^q + a_2 x^{q^2} + \ldots + a_{d-1} x^{q^{d-1}},$$

where $a_i \in \mathbb{K}$. Thus a linearised polynomial $f(x)$ is a polynomial whose evaluation $f(a)$ for any $a \in \mathbb{K}$ gives a linear combination of the d conjugates of a.

Linearised polynomials are linear transformations on $\mathbb{K}$, when considered as a vector space over $\mathbb{F}$. Conversely, any linear transformation of $\mathbb{K}$ over $\mathbb{F}$ can be expressed as a linearised polynomial.

EXAMPLE 2.71 Any linear transformation of $GF(2^8)$ as a vector space over $GF(2)$ can be represented by a (linearised) polynomial of the form $f(x) = a_0 x^{2^0} + a_1 x^{2^1} + a_2 x^{2^2} + \ldots + a_7 x^{2^7}$, where $a_i \in GF(2^8)$. □

We now consider the field $GF(p^d)$ as an extension field of $GF(p)$, where p is prime. The mapping $\tau \colon GF(p^d) \to GF(p^d)$ defined by $a \mapsto a^p$ maps a to one of its conjugates with respect to $GF(p)$. This mapping satisfies

$$\tau(a + a') = \tau(a) + \tau(a') \quad \text{and} \quad \tau(aa') = \tau(a)\tau(a').$$

Thus τ is a field automorphism of $GF(p^d)$, known as the *Frobenius* automorphism. The set of all automorphisms of $GF(p^d)$ under the operation of composition is the cyclic group of order d generated by τ. We note that τ fixes all elements of the subfield $GF(p)$ of $GF(p^d)$. Thus the automorphisms of $GF(p^d)$ are also linear transformations over $GF(p)$.

5. Varieties and Gröbner Bases

A large part of this monograph is concerned with expressing an AES encryption as a system of polynomial equations and considering methods of solution for such equations. In this section, we give a brief overview of the basic concepts used to analyse such equation systems.

Varieties

An *affine subset* of a vector space V is a coset or translate $U + u$ of some subspace $U < V$. The *affine space* based on V is the geometrical space given by considering certain geometrical properties of the affine subsets of V [58]. Thus we can usually identify the n-dimensional affine space over a field $\mathbb{F}$ with $\mathbb{F}^n$. The projective space $PG(n, \mathbb{F})$ is the geometrical space given by considering the one-dimensional subspaces of the $(n + 1)$-dimensional vector space $\mathbb{F}^{n+1}$. Thus we can represent an element of $PG(n, \mathbb{F})$ by a nonzero vector $(a_0, a_1, \ldots, a_n) \in \mathbb{F}^{n+1}$, where all nonzero scalar multiples of $(a_0, a_1, \ldots, a_n)$ represent the same element of $PG(n, \mathbb{F})$.

DEFINITION 2.72 Let $\mathbb{F}$ be a field and $\mathbb{F}^n$ denote the n-dimensional affine space over $\mathbb{F}$, and suppose that $f_1, \ldots, f_m$ are polynomials in $\mathbb{F}[x_1, \ldots, x_n]$. The *affine variety* defined by $f_1, \ldots, f_m$ is the subset of $\mathbb{F}^n$ given by

$$\{ (a_1, \ldots, a_n) \in \mathbb{F}^n \mid f_i(a_1, \ldots, a_n) = 0 \quad \text{for } i = 1, \ldots, m \}.$$

This variety is denoted by $\mathcal{V}(f_1, \ldots, f_m)$.

Thus the affine variety of Definition 2.72 describes the set of solutions in $\mathbb{F}$ of the polynomial equation system

$$f_1(x_1, \ldots, x_n) = 0, \quad \ldots \quad, f_m(x_1, \ldots, x_n) = 0.$$

EXAMPLE 2.73 Consider the polynomial ring $\mathbb{R}[x, y]$ in two variables, and let $f(x, y) = x^2 + y^2 - 1$ and $g(x, y) = x - 1$ be two polynomials in $\mathbb{R}[x, y]$. The affine variety $\mathcal{V}(f)$ consists of the points in the circle of radius 1 in $\mathbb{R}^2$ and is the solution set of the equation $x^2 + y^2 = 1$. The affine variety $\mathcal{V}(f, g) = \{(1, 0)\} \in \mathbb{R}^2$ is the set of solutions to $f(x, y) = g(x, y) = 0$. □

DEFINITION 2.74 Let $\mathrm{PG}(n, \mathbb{F})$ denote the projective space of dimension n. Suppose that $f_1, \ldots, f_m$ are homogeneous polynomials in the polynomial ring $\mathbb{F}[x_0, x_1, \ldots, x_n]$. The *projective variety* defined by the polynomials f_1, ..., f_m is the subset of $\mathrm{PG}(n, \mathbb{F})$ given by

$$\{ (a_0, a_1, \ldots, a_n) \in \mathrm{PG}(n, \mathbb{F}) \mid f_i(a_0, a_1, \ldots, a_n) = 0 \text{ for } i = 1, \ldots, m \}.$$

The projective space $\mathrm{PG}(n, \mathbb{F})$ can be partitioned into two subsets U and H, where

$$U = \{ (a_0, a_1, \ldots, a_n) \in \mathrm{PG}(n, \mathbb{F}) \mid a_0 \neq 0 \}, \text{ and}$$
$$H = \{ (0, a_1, \ldots, a_n) \in \mathrm{PG}(n, \mathbb{F}) \}.$$

The subset U can be identified with the affine space $\mathbb{F}^n$ by using the mapping

$$(a_0, a_1, \ldots, a_n) \mapsto \left(\frac{a_1}{a_0}, \ldots, \frac{a_n}{a_0} \right).$$

Furthermore, the subset H can be identified with the projective space $\mathrm{PG}(n - 1, \mathbb{F})$ by using the mapping $(0, a_1, \ldots, a_n) \mapsto (a_1, \ldots, a_n)$. Thus the projective space $\mathrm{PG}(n, \mathbb{F})$ can be partitioned into an affine space U and a projective space H of smaller dimension. The projective part H is known as the *hyperplane at infinity* of $\mathrm{PG}(n, \mathbb{F})$.

Given a projective variety $\mathcal{W} \in \mathrm{PG}(n, \mathbb{F})$, the set $\mathcal{V} = \mathcal{W} \cap U$ can be considered as an affine variety of $\mathbb{F}^n$ and is called the *affine portion* of $\mathcal{W}$. Thus every projective variety $\mathcal{W}$ can be seen as consisting of an affine variety $\mathcal{V}$ together with its points at infinity $\mathcal{W} \cap H$. Theorem 2.75 summarises the relationship between an affine and a projective variety.

THEOREM 2.75 Let $\mathcal{V} \subset \mathbb{F}^n$ be the affine variety defined by the polynomials $f_1, \ldots, f_m \in \mathbb{F}[x_1, \ldots, x_n]$. If f_i^h denotes the homogenisation of the polynomial f_i, then the variety $\mathcal{W}$ defined by the polynomials $f_1^h, \ldots, f_m^h \in \mathbb{F}[x_0, x_1, \ldots, x_n]$ is a projective variety of $\mathrm{PG}(n, \mathbb{F})$, of which the affine portion is $\mathcal{W} \cap U = \mathcal{V}$.

The above definitions of affine and projective varieties are given in terms of a finite set of polynomials. However, Theorem 2.76 shows that varieties are in fact defined by polynomial ideals.

THEOREM 2.76 Let I be an ideal of $\mathbb{F}[x_1, \ldots, x_n]$. If $\mathcal{V}(I)$ denotes the set

$$\{ (a_1, \ldots, a_n) \in \mathbb{F}^n \mid f(a_1, \ldots, a_n) = 0 \text{ for } f \in I \},$$

then $\mathcal{V}(I)$ is an affine variety. Furthermore, if $I = \langle f_1, \ldots, f_m \rangle$, then $\mathcal{V}(I) = \mathcal{V}(f_1, \ldots, f_m)$.

Similarly, a projective variety can be defined by a *homogeneous ideal* of $\mathbb{F}[x_0, x_1, \ldots, x_n]$, that is an ideal which is generated by homogeneous polynomials.

Gröbner bases

Theorem 2.76 means that the problem of finding the solutions of a polynomial equation system over a field $\mathbb{F}$ is often studied in the context of commutative algebra and polynomial ideals. The solution set of a particular system

$$f_1(x_1, \ldots, x_n) = 0, \ \ldots \ , f_m(x_1, \ldots, x_n) = 0$$

can be found by computing the variety $\mathcal{V}(I)$, where $I = \langle f_1, \ldots, f_m \rangle$. In particular, any generating set of I can be used to compute $\mathcal{V}(I)$. The *Hilbert Basis Theorem* states that any ideal $I \lhd \mathbb{F}[x_1, \ldots, x_n]$ is finitely generated. A Gröbner basis of the polynomial ideal I is a particular type of generating set of I and can be particularly useful in obtaining various properties of I, including the variety $\mathcal{V}(I)$.

DEFINITION 2.77 Suppose that $\mathbb{F}[x_1, \ldots, x_n]$ is a polynomial ring over a field $\mathbb{F}$ with a monomial ordering and that $I \lhd \mathbb{F}[x_1, \ldots, x_n]$ is a non–trivial ideal. We let $\mathrm{LT}(I)$ denote the set of all leading terms of elements of I and $\langle \mathrm{LT}(I) \rangle$ denote the ideal generated by the monomials in $\mathrm{LT}(I)$. A finite set $G = \{g_1, \ldots, g_s\} \subset I$ is said to be a *Gröbner basis* of I if

$$\langle \mathrm{LT}(g_1), \ldots, \mathrm{LT}(g_s) \rangle = \langle \mathrm{LT}(I) \rangle.$$

Thus G is a Gröbner basis of I if and only if the leading term of any polynomial in I is divisible by at least one of the leading terms $\{\mathrm{LT}(g_1), \ldots, \mathrm{LT}(g_s)\}$.

Every non–trivial ideal $I \lhd \mathbb{F}[x_1, \ldots, x_n]$ has a Gröbner basis, which is a generating set or basis for the ideal I. If G is a Gröbner basis of I

and $f \in I$, then the set $G \cup \{f\}$ satisfies Definition 2.77 and is also a Gröbner basis of I. Thus an ideal does not have a unique Gröbner basis.

DEFINITION 2.78 A *reduced Gröbner basis* for I is a Gröbner basis G such that the leading coefficient of every polynomial in G is 1 and none of the monomials of any $f \in G$ is divisible by the leading term of any other polynomial in G. Thus in a reduced Gröbner basis G, no monomial of $f \in G$ belongs to the ideal $\langle LT(G \setminus \{f\}) \rangle$.

Every non–trivial ideal I of $\mathbb{F}[x_1, \ldots, x_n]$ has a unique reduced Gröbner basis (with respect to a specific monomial ordering). We can obtain the reduced Gröbner basis for I from a Gröbner basis G for I by dividing or reducing each $f \in G$ by the set $G \setminus \{f\}$.

EXAMPLE 2.79 We consider the ring of real polynomials in three variables $\mathbb{R}[x, y, z]$ with the *grevlex* ordering. The set

$$\{z^6 - x^2 y, \ yz^4 + x, \ xy^2 + z^2\}$$

is a (reduced) Gröbner basis for the ideal of $\mathbb{R}[x, y, z]$ generated by these three polynomials. By contrast, consider the set

$$G = \{xy^2 + zx, \ y^2 z + z^2 - y\}$$

and the ideal I generated by these two polynomials. We have

$$xy = z(xy^2 + zx) - x(y^2 z + z^2 - y),$$

so $xy \in I$. However, xy is not divisible by the leading term of either polynomial in G (xy^2 or $y^2 z$). Thus G is not a Gröbner basis for the ideal I. □

Theorem 2.80 gives a sufficient condition in terms of the *greatest common divisor* of pairs of leading monomials for identifying whether a set is a Gröbner basis of a polynomial ideal.

THEOREM 2.80 Suppose $G \subset \mathbb{F}[x_1, \ldots, x_n]$ is a set of polynomials such that $\gcd(\mathrm{LM}(f), \mathrm{LM}(g)) = 1$ for all distinct $f, g \in G$. Then G is a Gröbner basis for the ideal $\langle G \rangle$.

Thus, if the leading monomials of all polynomials in a set G are pairwise coprime, then G is a Gröbner basis for the ideal generated by the polynomials of G. However, Example 2.79 shows that the condition of Theorem 2.80 is not necessary for a set G to be a Gröbner basis of $\langle G \rangle$.

Gröbner bases are an extremely powerful concept, with many applications in commutative algebra, algebraic geometry, and computational algebra. For example, Gröbner bases can be used to solve the ideal membership problem, that is to decide whether a polynomial f is in an ideal $I \lhd \mathbb{F}[x_1, \ldots, x_n]$. A polynomial f is in I if and only if f reduces to zero with respect to a Gröbner basis of I, that is the division of f by a Gröbner basis of I has remainder zero (Theorem 2.37).

The main relevance of Gröbner bases to cryptology is the problem of solving polynomial equation systems. If we have such a system

$$f_1(x_1, \ldots, x_n) = 0, \ldots, f_m(x_1, \ldots, x_n) = 0,$$

then we can find its solution set by computing the Gröbner basis for the ideal $I = \langle f_1, \ldots, f_m \rangle$ and computing the associated variety $\mathcal{V}(I)$. The Gröbner basis of I provides implicit solutions to the equation system over the algebraic closure of the field $\mathbb{F}$. A particularly useful monomial ordering for finding solutions to this polynomial equation system in $\mathbb{F}$ is the *lex* ordering, which is an example of an *elimination ordering*.

It is worth noting that equation systems arising in cryptography often display many properties. Cryptographic equation systems are often defined over a small finite field $\mathrm{GF}(q)$ and the solutions of cryptographic interest lie in this field. In this case, we could add the field relations $x_i^q - x_i$ to the original equation system. In this way the solutions of the extended equation system are restricted to the base field $\mathrm{GF}(q)$. Furthermore, cryptographic equation systems often have a unique solution $(a_1, \ldots, a_n) \in \mathrm{GF}(q)^n$. In this case, the reduced Gröbner basis of the ideal corresponding to the extended equation system would be $\{x_1 - a_1, \ldots, x_n - a_n\}$.

We discuss some methods and algorithms for computing a Gröbner basis of an ideal $I \lhd \mathbb{F}[x_1, \ldots, x_n]$ in Section 6.1.

Chapter 3

DESCRIPTION OF THE AES

This chapter gives a brief description of the AES and its design rationale. We place a particular emphasis on areas that are most relevant to subsequent chapters. The AES is a block cipher with a block size of 128 bits and a key size of 128, 192, or 256 bits. We denote these versions by AES-128, AES-192, and AES-256 respectively. This monograph focuses on AES-128 and we follow the formal description given in FIPS 197 [95].

1. Structure

The standard view of the AES is as a series of operations on a square array of 16 bytes [37, 39, 95]. The mathematical foundations that we need for this description are given in Chapter 2.

AES byte structure

An important aspect of any cipher is the method of data representation. We first discuss the structure used to represent a byte of data and then the structure used to represent the 16-byte blocks of data.

A byte is conventionally viewed as an ordered sequence of eight bits. Thus a byte consisting of the eight bits $b_7b_6b_5b_4b_3b_2b_1b_0$ can be viewed as a vector in an 8-dimensional vector space over $GF(2)$.

A byte can also be viewed as an element of the finite field $GF(2^8)$. The AES standard [95] specifies a representation of a byte in $GF(2^8)$ by defining this field in terms of the polynomial

$$m(x) = x^8 + x^4 + x^3 + x + 1,$$

which is irreducible in $GF(2)[x]$. We term the above polynomial $m(x)$ the *Rijndael polynomial*. From Section 2.4, there are two equivalent methods to define the field $GF(2^8)$ with respect to this irreducible polynomial,

$\longleftarrow$ column j $\longrightarrow$

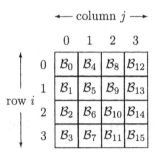

Figure 3.1. The AES array of bytes.

either as a quotient ring or as an extension field. We refer to the field $\mathrm{GF}(2^8)$ defined by the Rijndael polynomial as the *Rijndael field* and denote it by $\mathbf{F}$ throughout this monograph. Thus $\mathbf{F} = \mathrm{GF}(2)[x]/\langle m(x)\rangle$ or $\mathbf{F} = \mathrm{GF}(2)(\theta)$, where θ denotes a root of the Rijndael polynomial, termed the *Rijndael root*. The representation of a byte $\mathsf{b_7b_6b_5b_4b_3b_2b_1b_0}$ in $\mathbf{F}$ can then be given in either of the following two equivalent ways.

- Quotient Ring: $\mathsf{b_7}x^7 + \mathsf{b_6}x^6 + \mathsf{b_5}x^5 + \mathsf{b_4}x^4 + \mathsf{b_3}x^3 + \mathsf{b_2}x^2 + \mathsf{b_1}x + \mathsf{b_0}$.

- Extension Field: $\mathsf{b_7}\theta^7 + \mathsf{b_6}\theta^6 + \mathsf{b_5}\theta^5 + \mathsf{b_4}\theta^4 + \mathsf{b_3}\theta^3 + \mathsf{b_2}\theta^2 + \mathsf{b_1}\theta + \mathsf{b_0}$.

In the AES, bytes are represented as elements of the Rijndael field $\mathbf{F}$ and are combined using addition (which is equivalent to bitwise XOR) and multiplication in the field.

We use the common practice of representing a byte using hexadecimal notation, and we interpret such hexadecimal notation as a vector or field element depending on the context. For example, **24** represents the bit string 00100100, the column vector $(0,0,1,0,0,1,0,0)^T$, or the element $\theta^5 + \theta^2$ in the extension field, according to the context.

AES block structure

The AES transforms the plaintext into the ciphertext via a sequence of intermediate 128-bit states. Full details of how the AES represents a string of 128 bits as a set of 16 bytes are explicitly given in [95]. The 16 bytes in the state or a round key can be represented by the string of bytes

$$\mathcal{B}_0\mathcal{B}_1\mathcal{B}_2\mathcal{B}_3\mathcal{B}_4\mathcal{B}_5\mathcal{B}_6\mathcal{B}_7\mathcal{B}_8\mathcal{B}_9\mathcal{B}_{10}\mathcal{B}_{11}\mathcal{B}_{12}\mathcal{B}_{13}\mathcal{B}_{14}\mathcal{B}_{15}.$$

An equivalent representation is as a 4×4 array $\mathcal{S}$ of these bytes, where $\mathcal{S}_{(i,j)} = \mathcal{B}_{4j+i}$ $(0 \le i, j \le 3)$, as shown in Figure 3.1.

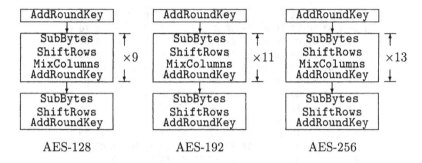

Figure 3.2. Schematic overview of the AES encryption.

Encryption

There are four basic operations when encrypting with the AES. These operate on the state array of 16 bytes.

- `SubBytes` modifies the bytes in the array independently.

- `ShiftRows` rotates the four rows of the array independently.

- `MixColumns` modifies the four columns of the array independently.

- `AddRoundKey` adds the bytes of the round key and the array.

These basic operations form a typical round of encryption. A complete description of AES encryption requires an initial `AddRoundKey` ("Round 0") followed by N_R rounds of computation, where $N_r = 10, 12,$ or 14 for AES-128, AES-196, or AES-256 respectively. The last round of computation does not contain a `MixColumns` operation. The sequence of operations for an AES encryption is summarised in Figure 3.2.

SubBytes.

The AES S-box $S[\cdot]$ provides a permutation of the set of 256 possible input bytes and is given as a look-up table in Figure 3.3. The operation `SubBytes` modifies the byte values by $\mathcal{B}_i \mapsto S[\mathcal{B}_i]$ $(0 \le i \le 15)$ or equivalently, the array elements $\mathcal{S}_{i,j}$ by $\mathcal{S}_{i,j} \mapsto S[\mathcal{S}_{i,j}]$ $(0 \le i, j \le 3)$. A mathematical description of `SubBytes` is given in Section 3.2.

ShiftRows.

Each row i $(0 \le i \le 3)$ of the 4×4 byte array S is rotated to the left by i positions. Thus `ShiftRows` modifies the byte values by (where subscripts are interpreted modulo 4)

$$\mathcal{S}_{i,j} \mapsto \mathcal{S}_{i,j-i} \ .$$

	-0	-1	-2	-3	-4	-5	-6	-7	-8	-9	-A	-B	-C	-D	-E	-F
0-	63	7C	77	7B	F2	6B	6F	C5	30	01	67	2B	FE	D7	AB	76
1-	CA	82	C9	7D	FA	59	47	F0	AD	D4	A2	AF	9C	A4	72	C0
2-	B7	FD	93	26	36	3F	F7	CC	34	A5	E5	F1	71	D8	31	15
3-	04	C7	23	C3	18	96	05	9A	07	12	80	E2	EB	27	B2	75
4-	09	83	2C	1A	1B	6E	5A	A0	52	3B	D6	B3	29	E3	2F	84
5-	53	D1	00	ED	20	FC	B1	5B	6A	CB	BE	39	4A	4C	58	CF
6-	D0	EF	AA	FB	43	4D	33	85	45	F9	02	7F	50	3C	9F	A8
7-	51	A3	40	8F	92	9D	38	F5	BC	B6	DA	21	10	FF	F3	D2
8-	CD	0C	13	EC	5F	97	44	17	C4	A7	7E	3D	64	5D	19	73
9-	60	81	4F	DC	22	2A	90	88	46	EE	B8	14	DE	5E	0B	DB
A-	E0	32	3A	0A	49	06	24	5C	C2	D3	AC	62	91	95	E4	79
B-	E7	C8	37	6D	8D	D5	4E	A9	6C	56	F4	EA	65	7A	AE	08
C-	BA	78	25	2E	1C	A6	B4	C6	E8	DD	74	1F	4B	BD	8B	8A
D-	70	3E	B5	66	48	03	F6	0E	61	35	57	B9	86	C1	1D	9E
E-	E1	F8	98	11	69	D9	8E	94	9B	1E	87	E9	CE	55	28	DF
F-	8C	A1	89	0D	BF	E6	42	68	41	99	2D	0F	B0	54	BB	16

Figure 3.3. The AES S-box used in SubBytes. The value of S[xy] is given at the intersection of row x- and column -y.

MixColumns.

Each column of the $4{\times}4$ byte array $\mathcal{S}$ is regarded as a column vector over the Rijndael field $\mathbf{F}$. It is then updated by multiplying the column vector by a specified $4{\times}4$ matrix over $\mathbf{F}$. Thus MixColumns modifies the state array by the matrix multiplication $(0 \le j \le 3)$

$$
\begin{pmatrix} \mathcal{S}_{0,j} \\ \mathcal{S}_{1,j} \\ \mathcal{S}_{2,j} \\ \mathcal{S}_{3,j} \end{pmatrix}
\mapsto
\begin{pmatrix} 02 & 03 & 01 & 01 \\ 01 & 02 & 03 & 01 \\ 01 & 01 & 02 & 03 \\ 03 & 01 & 01 & 02 \end{pmatrix}
\begin{pmatrix} \mathcal{S}_{0,j} \\ \mathcal{S}_{1,j} \\ \mathcal{S}_{2,j} \\ \mathcal{S}_{3,j} \end{pmatrix}.
$$

AddRoundKey.

The AES key schedule processes the user-supplied key to give the 16-byte round keys $\mathcal{K}_{r,0} \ldots \mathcal{K}_{r,15}$ $(0 \le r \le N_r)$ for the AES. In round r, AddRoundKey updates the state array by $\mathcal{B}_i \mapsto \mathcal{B}_i + \mathcal{K}_{r,i}$ $(0 \le i \le 15)$ or equivalently by $\mathcal{S}_{i,j} \mapsto \mathcal{S}_{i,j} + \mathcal{K}_{r,4j+i}$ $(0 \le i, j \le 3)$.

Key schedule

The generation of the AES round keys is straightforward even though three key sizes are supported. Generally speaking, key material is generated recursively, and at each round sufficient key material is extracted to form a 128-bit round key. We only give a description of the key schedule

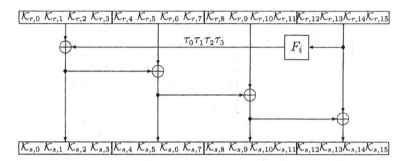

Figure 3.4. A schematic overview of the AES-128 key schedule.

for AES-128, though the key schedules for AES-192 and AES-256 are similar. Full details are given in [95].

We assume that the round key at round r $(0 \leq r \leq 10)$ is given by $\mathcal{K}_{r,0} \ldots \mathcal{K}_{r,15}$ where the user-supplied key forms the round key at round 0. In order to form the round key for round $s = r + 1$, we first define a temporary word $T_0 T_1 T_2 T_3$ of four bytes by

$$T_0 = S[\mathcal{K}_{r,13}] + \theta^r, T_1 = S[\mathcal{K}_{r,14}], T_2 = S[\mathcal{K}_{r,15}], \text{ and } T_3 = S[\mathcal{K}_{r,12}],$$

where θ is the Rijndael root. The key for round s is then given by

$$\mathcal{K}_{s,i} = \begin{cases} \mathcal{K}_{r,i} + T_i & 0 \leq i \leq 3 \\ \mathcal{K}_{r,i} + \mathcal{K}_{s,i-4} & 4 \leq i \leq 15. \end{cases}$$

In summary, the temporary word $T_0 T_1 T_2 T_3$ is generated using the non-linear key schedule function F_i. This consists of applying the S-box to all components of the input, a rotation of bytes, and the addition of a round-specific constant. This process is illustrated in Figure 3.4.

Decryption

Decryption for the AES can be performed by using the inverse of the four operations in reverse order, with the round keys taken in reverse order. Since the operations SubBytes and ShiftRows commute and MixColumns is omitted from the final round [37, 39, 95], there is an equivalent description of the AES decryption that mimics the sequence of operations during encryption.

InvSubBytes.

The inverse of the AES S-box, $S^{-1}[\cdot]$, is easily derived. The operation InvSubBytes modifies the state by $\mathcal{B}_i \mapsto S^{-1}[\mathcal{B}_i]$ $(0 \leq i \leq 15)$ or, in

	-0	-1	-2	-3	-4	-5	-6	-7	-8	-9	-A	-B	-C	-D	-E	-F
0-	52	09	6A	D5	30	36	A5	38	BF	40	A3	9E	81	F3	D7	FB
1-	7C	E3	39	82	9B	2F	FF	87	34	8E	43	44	C4	DE	E9	CB
2-	54	7B	94	32	A6	C2	23	3D	EE	4C	95	0B	42	FA	C3	4E
3-	08	2E	A1	66	28	D9	24	B2	76	5B	A2	49	6D	8B	D1	25
4-	72	F8	F6	64	86	68	98	16	D4	A4	5C	CC	5D	65	B6	92
5-	6C	70	48	50	FD	ED	B9	DA	5E	15	46	57	A7	8D	9D	84
6-	90	D8	AB	00	8C	BC	D3	0A	F7	E4	58	05	B8	B3	45	06
7-	D0	2C	1E	8F	CA	3F	0F	02	C1	AF	BD	03	01	13	8A	6B
8-	3A	91	11	41	4F	67	DC	EA	97	F2	CF	CE	F0	B4	E6	73
9-	96	AC	74	22	E7	AD	35	85	E2	F9	37	E8	1C	75	DF	6E
A-	47	F1	1A	71	1D	29	C5	89	6F	B7	62	0E	AA	18	BE	1B
B-	FC	56	3E	4B	C6	D2	79	20	9A	DB	C0	FE	78	CD	5A	F4
C-	1F	DD	A8	33	88	07	C7	31	B1	12	10	59	27	80	EC	5F
D-	60	51	7F	A9	19	B5	4A	0D	2D	E5	7A	9F	93	C9	9C	EF
E-	A0	E0	3B	4D	AE	2A	F5	B0	C8	EB	BB	3C	83	53	99	61
F-	17	2B	04	7E	BA	77	D6	26	E1	69	14	63	55	21	0C	7D

Figure 3.5. The inverse of the AES S-box used in `InvSubBytes`. The value of $S^{-1}[xy]$ is given at the intersection of row `x`- and column `-y`.

terms of the state array, the array elements $\mathcal{S}_{i,j}$ by $\mathcal{S}_{i,j} \mapsto S^{-1}[\mathcal{S}_{i,j}]$ ($0 \leq i, j \leq 3$). The inverse of the S-box is given as a look-up table in Figure 3.5.

`InvShiftRows`.

Each row i ($0 \leq i \leq 3$) of the 4×4 byte array $\mathcal{S}$ is rotated to the right by i positions. Thus `InvShiftRows` modifies the byte values by $\mathcal{S}_{i,j} \mapsto \mathcal{S}_{i,j+i}$, where the subscripts are interpreted modulo 4.

`InvMixColumns`.

Each column of the 4×4 byte array $\mathcal{S}$ is regarded as a column vector over the Rijndael field $\mathbf{F}$, which is updated using the inverse of the 4×4 `MixColumns` matrix over $\mathbf{F}$. Thus `InvMixColumns` modifies the state array by the matrix multiplication ($0 \leq j \leq 3$)

$$\begin{pmatrix} \mathcal{S}_{0,j} \\ \mathcal{S}_{1,j} \\ \mathcal{S}_{2,j} \\ \mathcal{S}_{3,j} \end{pmatrix} \mapsto \begin{pmatrix} 0E & 0B & 0D & 09 \\ 09 & 0E & 0B & 0D \\ 0D & 09 & 0E & 0B \\ 0B & 0D & 09 & 0E \end{pmatrix} \begin{pmatrix} \mathcal{S}_{0,j} \\ \mathcal{S}_{1,j} \\ \mathcal{S}_{2,j} \\ \mathcal{S}_{3,j} \end{pmatrix}.$$

	-0	-1	-2	-3	-4	-5	-6	-7	-8	-9	-A	-B	-C	-D	-E	-F
0-	00	01	8D	F6	CB	52	7B	D1	E8	4F	29	C0	B0	E1	E5	C7
1-	74	B4	AA	4B	99	2B	60	5F	58	3F	FD	CC	FF	40	EE	B2
2-	3A	6E	5A	F1	55	4D	A8	C9	C1	0A	98	15	30	44	A2	C2
3-	2C	45	92	6C	F3	39	66	42	F2	35	20	6F	77	BB	59	19
4-	1D	FE	37	67	2D	31	F5	69	A7	64	AB	13	54	25	E9	09
5-	ED	5C	05	CA	4C	24	87	BF	18	3E	22	F0	51	EC	61	17
6-	16	5E	AF	D3	49	A6	36	43	F4	47	91	DF	33	93	21	3B
7-	79	B7	97	85	10	B5	BA	3C	B6	70	D0	06	A1	FA	81	82
8-	83	7E	7F	80	96	73	BE	56	9B	9E	95	D9	F7	02	B9	A4
9-	DE	6A	32	6D	D8	8A	84	72	2A	14	9F	88	F9	DC	89	9A
A-	FB	7C	2E	C3	8F	B8	65	48	26	C8	12	4A	CE	E7	D2	62
B-	0C	E0	1F	EF	11	75	78	71	A5	8E	76	3D	BD	BC	86	57
C-	0B	28	2F	A3	DA	D4	E4	0F	A9	27	53	04	1B	FC	AC	E6
D-	7A	07	AE	63	C5	DB	E2	EA	94	8B	C4	D5	9D	F8	90	6B
E-	B1	0D	D6	EB	C6	0E	CF	AD	08	4E	D7	E3	5D	50	1E	B3
F-	5B	23	38	34	68	46	03	8C	DD	9C	7D	A0	CD	1A	41	1C

Figure 3.6. The AES inversion within the S-box.

2. Design Rationale

Each component of the AES was carefully chosen and has a specific role. The design rationale is discussed in [37, 39] and we only present the essential points here. Each round of the AES is considered to have three parts. The first is `SubBytes`, in which a substitution is performed on each byte of the state array. This is termed the *substitution layer*. The second part is `ShiftRows` followed by `MixColumns`, which gives diffusion across the state array. This is termed the *diffusion layer*. The final part of an AES round introduces key material by `AddRoundKey`. We now discuss the substitution and diffusion layers.

Substitution layer

The substitution layer is based on the AES S-box which is, in turn, defined by the composition of three operations.

- *Inversion*. The AES *inversion* operation is inversion in the Rijndael field $\mathbf{F}$, but extended so that $0 \mapsto 0$. Thus, the input byte to the S-box is regarded as an element $w \in \mathbf{F}$ and for $w \neq 0$ the output x satisfies $x = w^{-1}$ and $wx = 1$. We denote the extension to the case $w = 0$ by $x = w^{(-1)}$ and give a look-up table in Figure 3.6.

- *GF(2)-linear mapping*. The GF(2)-linear mapping is a linear transformation $\xi : \text{GF}(2)^8 \to \text{GF}(2)^8$ specified by an 8×8 circulant matrix over GF(2). The result x of *inversion* is regarded as a vector in

	-0	-1	-2	-3	-4	-5	-6	-7	-8	-9	-A	-B	-C	-D	-E	-F
0-	00	1F	3E	21	7C	63	42	5D	F8	E7	C6	D9	84	9B	BA	A5
1-	F1	EE	CF	D0	8D	92	B3	AC	09	16	37	28	75	6A	4B	54
2-	E3	FC	DD	C2	9F	80	A1	BE	1B	04	25	3A	67	78	59	46
3-	12	0D	2C	33	6E	71	50	4F	EA	F5	D4	CB	96	89	A8	B7
4-	C7	D8	F9	E6	BB	A4	85	9A	3F	20	01	1E	43	5C	7D	62
5-	36	29	08	17	4A	55	74	6B	CE	D1	F0	EF	B2	AD	8C	93
6-	24	3B	1A	05	58	47	66	79	DC	C3	E2	FD	A0	BF	9E	81
7-	D5	CA	EB	F4	A9	B6	97	88	2D	32	13	0C	51	4E	6F	70
8-	8F	90	B1	AE	F3	EC	CD	D2	77	68	49	56	0B	14	35	2A
9-	7E	61	40	5F	02	1D	3C	23	86	99	B8	A7	FA	E5	C4	DB
A-	6C	73	52	4D	10	0F	2E	31	94	8B	AA	B5	E8	F7	D6	C9
B-	9D	82	A3	BC	E1	FE	DF	C0	65	7A	5B	44	19	06	27	38
C-	48	57	76	69	34	2B	0A	15	B0	AF	8E	91	CC	D3	F2	ED
D-	B9	A6	87	98	C5	DA	FB	E4	41	5E	7F	60	3D	22	03	1C
E-	AB	B4	95	8A	D7	C8	E9	F6	53	4C	6D	72	2F	30	11	0E
F-	5A	45	64	7B	26	39	18	07	A2	BD	9C	83	DE	C1	E0	FF

Figure 3.7. The AES GF(2)-linear mapping within the S-box.

$GF(2)^8$, and the output vector y is given by $y = \xi(x)$, where

$$
\begin{pmatrix} y_7 \\ y_6 \\ y_5 \\ y_4 \\ y_3 \\ y_2 \\ y_1 \\ y_0 \end{pmatrix}
=
\begin{pmatrix}
1 & 0 & 0 & 0 & 1 & 1 & 1 & 1 \\
1 & 1 & 0 & 0 & 0 & 1 & 1 & 1 \\
1 & 1 & 1 & 0 & 0 & 0 & 1 & 1 \\
1 & 1 & 1 & 1 & 0 & 0 & 0 & 1 \\
1 & 1 & 1 & 1 & 1 & 0 & 0 & 0 \\
0 & 1 & 1 & 1 & 1 & 1 & 0 & 0 \\
0 & 0 & 1 & 1 & 1 & 1 & 1 & 0 \\
0 & 0 & 0 & 1 & 1 & 1 & 1 & 1
\end{pmatrix}
\begin{pmatrix} x_7 \\ x_6 \\ x_5 \\ x_4 \\ x_3 \\ x_2 \\ x_1 \\ x_0 \end{pmatrix}.
$$

We give a look-up table for the GF(2)-linear mapping in Figure 3.7.

- *S-box constant.* The output byte y of the GF(2)-linear mapping is regarded as an element of the Rijndael field **F** and added to the field element **63** to produce the output from the S-box.

The rationale for using the inversion operation is that it provides good local resistance [98, 99] to the standard block cipher cryptanalytic techniques of differential [11, 10, 68] and linear [78] cryptanalysis. The rationale for the use of the GF(2)-linear mapping and the S-box constant is to increase the algebraic complexity of the S-box and to remove fixed points respectively [37, 39].

Diffusion layer

The diffusion layer has been designed in accordance with the *wide trail strategy* [35, 39]. For the AES, the wide trail strategy is based on the 4×4 matrix over **F** used in MixColumns. This matrix is the parity check matrix for a *maximal distance separable* (*MDS*) code [76], and such a matrix is known as an *MDS matrix* (Definition 2.52). A 4×4 matrix over **F** operates on four input bytes and gives four output bytes. For the 4×4 MDS matrix used in MixColumns, either all the input and output bytes are zero, or at least five of these eight bytes are non-zero. This MDS property is used to ensure that the number of *active* S-boxes involved in a differential or linear attack increases rapidly, and the security of the AES against these particular attacks can be established.

0-inversion

When 00 is used as input to an AES S-box, inversion in the Rijndael field **F** is extended, and we term this a 0-*inversion*. The AES-128 has 10 rounds and each round requires 16 S-box computations, so the probability of there being no 0-inversions during an AES-128 encryption is $(\frac{255}{256})^{160} \approx 0.53$. Similarly, the AES-128 key schedule requires 40 S-box computations, so the probability of there being no 0-inversions during an AES key setup is $(\frac{255}{256})^{40} \approx 0.86$. These calculations assume statistical independence of the 0-inversions.

3. Small Scale Variants of the AES

In this section we describe small scale variants of the AES that are intended to provide a fully parameterised framework for detailed analysis [22]. Other small scale variants have been proposed, though usually as an educational, rather than an experimental, tool [91, 102].

Two sets of small scale variants of the AES are defined in [22] and these differ only in the form of the final round. The two sets of variants are denoted $\text{SR}(r, n_R, n_C, e)$ and $\text{SR}^*(r, n_R, n_C, e)$, with $\text{SR}(r, n_R, n_C, e)$ including a MixColumns operation in the last round. Both are parameterised in the following way:

- r is the number of rounds,

- n_R is the number of rows in the rectangular grid of the state,

- n_C is the number of columns in the rectangular grid of the state,

- e is the word size (in bits).

Both $\text{SR}(r, n_R, n_C, e)$ and $\text{SR}^*(r, n_R, n_C, e)$ have a block size of $n_R n_C e$ bits and the full AES is modelled by $\text{SR}^*(10, 4, 4, 8)$. The data block is

Inversion in $\mathrm{GF}(2^4)$

Input	0	1	2	3	4	5	6	7	8	9	A	B	C	D	E	F
Output	0	1	9	E	D	B	7	6	F	2	C	5	A	4	3	8

$\mathrm{GF}(2)$-linear mapping in $\mathrm{GF}(2^4)$

Input	0	1	2	3	4	5	6	7	8	9	A	B	C	D	E	F
Output	0	D	B	6	7	A	C	1	E	3	5	8	9	4	2	F

Full S-box over $\mathrm{GF}(2^4)$ with S-box constant 6

Input	0	1	2	3	4	5	6	7	8	9	A	B	C	D	E	F
Output	6	B	5	4	2	E	7	A	9	D	F	C	3	1	0	8

Figure 3.8. An equivalent S-box over $\mathrm{GF}(2^4)$ for small scale variants of the AES.

viewed as an $n_R \times n_C$ array of words of e bits. Useful small scale variants exist when both n_R and n_C are restricted to 1, 2, or 4. Examples of such arrays with words numbered in the AES style are given below.

| 0 |

0
1

0	2
1	3

0	4
1	5
2	6
3	7

0	2	4	6
1	3	5	7

The word sizes $e = 4$ and $e = 8$ are the most relevant and are defined with respect to the fields $\mathrm{GF}(2^4)$ and $\mathrm{GF}(2^8)$. The field $\mathrm{GF}(2^4)$ is defined by the primitive polynomial $x^4 + x + 1$ over $\mathrm{GF}(2)$ with root ρ. Thus $\mathrm{SR}(n, c, e, 4)$ uses the field $\mathrm{GF}(2)[X]/\langle x^4 + x + 1\rangle$ or equivalently $\mathrm{GF}(2)(\rho)$. Small scale variants over $\mathrm{GF}(2^8)$ use the Rijndael field $\mathbf{F}$.

We define a round of the small scale variants over the field $\mathrm{GF}(2^4)$ by describing variants of the AES operations. An S-box over $\mathrm{GF}(2^4)$ consists of the following three (sequential) operations, which are summarised in Figure 3.8.

- *Inversion.* The first operation is an extended inversion in the field $\mathrm{GF}(2^4)$ (with $0 \mapsto 0$).

- *$\mathrm{GF}(2)$-linear mapping.* The 4×4 matrix over $\mathrm{GF}(2)$ used to define the $\mathrm{GF}(2)$-linear matrix is the circulant matrix

$$\begin{pmatrix} 1 & 1 & 1 & 0 \\ 0 & 1 & 1 & 1 \\ 1 & 0 & 1 & 1 \\ 1 & 1 & 0 & 1 \end{pmatrix}.$$

- *S-box constant.* The output from the S-box is produced by adding the S-box constant 6 to the output of the $\mathrm{GF}(2)$-linear mapping.

The small scale equivalent of `ShiftRows` is the simultaneous left rotation of row i in the data array by i positions $(0 \leq i \leq n_R - 1)$. The small scale equivalent of `MixColumns` multiplies each column of the state array by an invertible circulant MDS matrix over $\mathrm{GF}(2^e)$. The matrices required for the different variants are all specified in [22], and they preserve the essential qualities of the original AES operation. Finally, the small scale variant of `AddRoundKey` is the obvious analogue, with the corresponding key schedules also being defined in [22].

These small scale variants retain, as far as possible, the algebraic features of the AES. They often have a small key space and can easily be analysed by exhaustive key search or equivalent techniques. However, the main purpose of these small scale variants is to assist in the algebraic analysis of the AES. Some experimental results based on these small scale variants are discussed in Chapter 6.

Chapter 4

ALGEBRAIC PROPERTIES OF THE AES

The first public comments on the algebraic structure of Rijndael were made towards the end of the AES selection process [67, 88, 111]. This chapter provides a summary of much of the related work that has followed the publication of Rijndael as the AES.

1. Round Structure

The AES design is an example of an SP-network, in which each round usually consists of three phases [113]. The first phase is a localised nonlinear transformation or substitution of the state, that is nonlinear transformations are applied to the various sub-blocks of the state. The second phase is an extensive linear diffusion of the entire state. The final phase combines the state with the key material. In the design rationale for the AES, the first phase is performed by the SubBytes operation, the second phase by the combination of the ShiftRows and MixColumns operations, and the final phase by the AddRoundKey operation.

The role of ShiftRows and MixColumns is to provide diffusion within the AES. They are both linear transformations of the cipher state over the Rijndael field **F**. The ShiftRows operation provides what is termed *high dispersion*, whilst the MixColumns operation provides *high local diffusion*. The two operations combine to give a highly efficient diffusion as required in the wide trail strategy [39]. However, in this section we give an alternative method of analysis [38, 87, 88] for diffusion in the AES. This analysis uses simple algebraic tools to explore the underlying structure of the AES component operations and their combination. We begin by considering the operations in a single round of the AES.

SubBytes operation

The AES S-box has three component transformations, namely the augmented inversion $w \to w^{(-1)}$, a GF(2)-linear mapping, and the addition of a constant (Section 3.2). The inversion operation has properties [98] that resist standard cryptanalysis, while the other components in the S-box are used to disguise its algebraic simplicity and to provide a "complicated algebraic expression if combined with the inverse mapping" [37]. In this way, an argument can be made for the resistance of the AES to the interpolation and similar attacks [62, 63]. Furthermore, the S-box constant 63 was "chosen in such a way that the S-box has no fixed points and no opposite fixed points" [37].

The final two S-box operations, the GF(2)-linear mapping and the addition of the S-box constant 63, form an affine transformation over GF(2). The 8×8 matrix for the GF(2)-linear mapping on a byte is given in Section 3.2. The GF(2)-linear mapping on the entire state space is thus given by a 128×128 matrix L over GF(2), where L is a block diagonal matrix with blocks given by this circulant 8×8 matrix.

ShiftRows and MixColumns operations

The ShiftRows operation is based on the rotation of rows of the state array. The rotation of a row by one position is represented by the 4×4 permutation matrix $\hat{R}$ over **F**, where

$$\hat{R} = \begin{pmatrix} 0 & 1 & 0 & 0 \\ 0 & 0 & 1 & 0 \\ 0 & 0 & 0 & 1 \\ 1 & 0 & 0 & 0 \end{pmatrix}.$$

If we change the basis of the state space so that the state array is represented by the column vector

$$(\mathcal{S}_0, \mathcal{S}_4, \mathcal{S}_8, \mathcal{S}_{12}, \mathcal{S}_1, \mathcal{S}_5, \mathcal{S}_9, \mathcal{S}_{13}, \mathcal{S}_2, \mathcal{S}_6, \mathcal{S}_{10}, \mathcal{S}_{14}, \mathcal{S}_3, \mathcal{S}_7, \mathcal{S}_{11}, \mathcal{S}_{15})^T,$$

then the action of the ShiftRows operation is represented by the 16×16 block diagonal matrix

$$\begin{pmatrix} I & 0 & 0 & 0 \\ 0 & \hat{R} & 0 & 0 \\ 0 & 0 & \hat{R}^2 & 0 \\ 0 & 0 & 0 & \hat{R}^3 \end{pmatrix}.$$

By re-ordering the rows and columns of this matrix, we can obtain a 16×16 matrix $\overline{R}$ over **F** that represents the ShiftRows operation with respect to the standard state array ordering by column.

The MixColumns operation is typically described in terms of the 4×4 MDS matrix over $\mathbf{F}$ given in Section 3.1. However, there is a basis of $\mathbf{F}^4$ such that the transformation given by the MDS matrix can be represented using the matrix $\hat{R}$. Thus there is a basis of $\mathbf{F}^{16}$ such that the MixColumns operation is given by the block diagonal matrix

$$\begin{pmatrix} \hat{R} & 0 & 0 & 0 \\ 0 & \hat{R} & 0 & 0 \\ 0 & 0 & \hat{R} & 0 \\ 0 & 0 & 0 & \hat{R} \end{pmatrix}.$$

The 16×16 matrix $\overline{C}$ that represents the MixColumns operation with respect to the standard state array ordering has the same algebraic properties as this matrix.

The combined action of ShiftRows followed by MixColumns is represented by the 16×16 matrix $\overline{C}\,\overline{R}$ over $\mathbf{F}$. This represents the linear diffusion provided by the wide trail strategy. Since both $\overline{C}$ and $\overline{R}$ are fundamentally based on $\hat{R}$, we can determine the simple algebraic properties of these constituent operations and their combination.

We can regard the state vector either as a vector over $\mathbf{F}$ of length 16 or as a vector over GF(2) of length 128. In the latter case, the ShiftRows and MixColumns operations are given by 128×128 matrices R and C respectively. Multiplication by an element of $\mathbf{F}$ is a linear transformation of $\mathbf{F}$ considered as the vector space GF(2)^8, and so multiplication is described by an 8×8 matrix over GF(2). Thus the matrices R and C are given by the block matrices in which the entries 1, θ and $\theta+1$ of $\overline{C}$ and $\overline{R}$ are replaced by the 8×8 matrices I, T_θ and $T_{\theta+1}$ (Example 2.65). The linear diffusion required by the wide trail strategy is therefore given by the 128×128 matrix CR over GF(2). The algebraic properties of R and C, and hence CR, are directly given by those of $\overline{R}$ and $\overline{C}$.

Augmented linear diffusion

We have seen that the final two parts of the SubBytes operation, namely the GF(2)-linear mapping and the addition of the S-box constant, form an affine operation over GF(2). Furthermore, the diffusion operations of the AES on bytes of the state space, namely ShiftRows and MixColumns, are also linear operations over GF(2). It is thus reasonable to consider an augmented linear diffusion for the AES, consisting of combining the affine transformation within the SubBytes operation with the ShiftRows and MixColumns operations. By combining these operations into one, we derive a very natural mathematical division of the AES round function. The first part consists of the simultaneous inversion of all components of the state. The second part is affine over $GF(2)$ and consists of the composition of all other operations in the round.

The division of the round function of a block cipher into a nonlinear part and an affine part is somewhat arbitrary, as there are clearly many ways in which such a division could be made. An algorithm for finding such decompositions is given in [12]. However, given that this nonlinear part is particularly simple, the division of the AES round function into a nonlinear inversion part and an affine part consisting of all other operations is strikingly clear.

We now give an expression for this affine second part of the AES round function, that is the round function without the inversion operation. We consider vectors of length 128 over $GF(2)$, and we suppose that $\mathbf{x}$ is the output of the inversion within the round, that $\mathbf{k}_i$ is the round key, and that **63** is the vector of repeated S-box constants. The affine part of the AES round mapping is then given by

$$\mathbf{x} \mapsto CR(L\mathbf{x} + \mathbf{63}) + \mathbf{k}_i,$$

where C, R, and L are the matrices discussed above. However, since $C(\mathbf{63}) = R(\mathbf{63}) = \mathbf{63}$, this affine mapping is given by

$$\mathbf{x} \mapsto CRL\mathbf{x} + \mathbf{k}_i + \mathbf{63}.$$

The linear transformation of this affine mapping is thus given by the $128{\times}128$ matrix $M = CRL$ over $GF(2)$, and this matrix is given in Appendix B. The matrix M is particularly simple and is only slightly more complicated than the linear diffusion matrix CR identified by the wide trail strategy. We therefore consider the *augmented linear diffusion* given by the matrix M in our subsequent analysis of the AES. This allows us to express the round function of the AES succinctly.

We now give such an expression for the AES round function. We suppose that $\mathbf{w}$ and $\mathbf{x}$ are vectors over $GF(2)$ of length 128, and that $\mathbf{w}$ is the input and $\mathbf{x}$ the output of the inversion operation. We then have

$$\mathbf{x} = (x_0, \ldots, x_{15})^T = \left(w_0^{(-1)}, \ldots, w_{15}^{(-1)} \right)^T = \mathbf{w}^{(-1)},$$

where $\mathbf{w}^{(-1)}$ denotes component-wise inversion. Furthermore, we can define a revised key schedule for the AES with round keys given by $\mathbf{k}_i^* = \mathbf{k}_i + \mathbf{63}$ $(i > 0)$ with $\mathbf{k}_0^* = \mathbf{k}_0$. A round of the AES is then given by

$$\mathbf{w} \mapsto M\mathbf{w}^{(-1)} + \mathbf{k}_i^*.$$

We can therefore consider an equivalent definition of the AES round function in which an S-box consists solely of the inversion operation.

	Matrix used in the Augmented Diffusion				
	C	R	CR	L	M
Minimal Polynomial	$x^4 + 1$	$x^4 + 1$	$x^8 + 1$	$(x+1)^3$	$(x+1)^{15}$
Order	4	4	8	4	16
Dimension: Fixed Subspace	32	64	16	48	16
Dimension: Order 2 Subspace	64	96	32	96	30
Dimension: Order 4 Subspace	128	128	64	128	58
Dimension: Order 8 Subspace	128	128	128	128	96

Figure 4.1. Some properties of matrices used in the augmented diffusion of the AES. The order i subspace for the matrix T is $\{\mathbf{v}|T^i\mathbf{v} = \mathbf{v}\}$.

Thus, while a design criterion for the S-box is that there be no "fixed points" [37], the equivalent S-box in the algebraically simpler description of the AES has two fixed points (00 and 01). The diffusion in the AES round is now given by the augmented linear diffusion, and a round of the AES consists solely of the following two simple algebraic operations.

- A component-wise inversion over the field $\mathbf{F}$.

- An affine transformation of a vector space over $\mathrm{GF}(2)$.

Properties of the augmented linear diffusion

We now discuss some of the basic properties of the augmented linear diffusion matrix $M = CRL$ and its component matrices. We summarise these properties in Figure 4.1. In particular, the minimal polynomial $\mathrm{min}_M(x)$ of M is given by

$$\mathrm{min}_M(x) = (x+1)^{15},$$

so $\mathrm{min}_M(x)$ divides $(x+1)^{16} = x^{16} + 1$. Thus M has order 16. This means that any 128-bit input to the augmented linear diffusion of the AES will be mapped to itself after at most 16 repeated applications of the augmented linear diffusion. Such a small order is notable because it suggests that the augmented linear diffusion possesses considerable structure, even though it includes two of the three parts of the highly nonlinear S-box. Furthermore, the affine transformation $A_{\mathbf{k}}$ of the augmented diffusion given by $\mathbf{x} \mapsto M\mathbf{x} + \mathbf{k}$ has order 16 since

$$A_{\mathbf{k}}^{16}\mathbf{x} = M^{16}\mathbf{x} + (M^{15} + M^{14} + \ldots + M + I)\,\mathbf{k}$$
$$= I\,\mathbf{x} + (M+I)^{15}\mathbf{k} = \mathbf{x}.$$

Further properties of the augmented linear diffusion matrix can be found by applying a change of basis transformation to the matrix M.

In particular, the augmented linear diffusion can be represented by the simple matrix $P^{-1}MP$ given in Appendix B. The simple structure of the augmented linear diffusion of the AES revealed by $P^{-1}MP$ gives 15 subspaces $V_1, \ldots, V_{15} \subseteq \mathrm{GF}(2)^{128}$ such that

$$\mathrm{GF}(2)^{128} = V_1 \oplus V_2 \oplus \ldots \oplus V_{15}.$$

These subspaces $V_1, \ldots, V_{15}$ have dimensions 16, 14, 14, 14, 10, 10, 10, 8, 8, 6, 4, 4, 4, 4, and 2, respectively. They have the property that if $\mathbf{v}_i \in V_i$, then $M\mathbf{v}_i = \mathbf{v}_i + \mathbf{v}_{i-1}$ for some $\mathbf{v}_{i-1} \in V_{i-1}$ $(i > 1)$ with $M\mathbf{v}_1 = \mathbf{v}_1$. Furthermore the subspaces $V_j' = V_1 \oplus \ldots \oplus V_j$ $(j = 1, \ldots, 15)$ are M-invariant, that is $MV_j' = V_j'$.

Such properties of the augmented linear diffusion of the AES suggest several ideas for analysis of the AES, some of which are mentioned below. The techniques are based on those given in [84, 86] and require some understanding of the structure and construction of these invariant subspaces [88].

We have seen that the 16-dimensional subspace $V_1 = V_1'$ is fixed by the matrix M. Thus there are 2^{16} vectors fixed by the augmented diffusion. Furthermore suppose that $\mathbf{x}$ and $\mathbf{x}'$ are two vectors such that the difference $\mathbf{x} + \mathbf{x}' \in V_1$, then

$$M\mathbf{x} + M\mathbf{x}' = M(\mathbf{x} + \mathbf{x}') = \mathbf{x} + \mathbf{x}',$$

and so the augmented diffusion of the AES also fixes 2^{16} differences. In particular, there exist vectors that are fixed by M and are nonzero for only 12 of the 16 bytes of the state. Thus the use of such a difference in an analysis of the AES would involve only 12 active S-boxes in each round. One such vector over $\mathrm{GF}(2)$ given in hexadecimal notation is

$$(55336600\ 33550066\ 55336600\ 33550066)^T.$$

Such an analysis of the augmented diffusion matrix M extends to parity checks. In this case, a *parity check* is a row vector $\mathbf{e}^T$ of length 128 over $\mathrm{GF}(2)$, and the *parity check value* of a vector $\mathbf{x}$ is $\mathbf{e}^T\mathbf{x} \in \mathrm{GF}(2)$. Furthermore, there are also 2^{16} row vectors $\mathbf{e}^T$ that are fixed by the augmented diffusion matrix M, that is $\mathbf{e}^T M = \mathbf{e}^T$. For such a parity check $\mathbf{e}^T$, any parity check value is fixed by the augmented diffusion as

$$\mathbf{e}^T M\mathbf{x} = \mathbf{e}^T\mathbf{x}.$$

Similarly, any parity check value of a difference is also fixed by the augmented diffusion as

$$\mathbf{e}^T (M\mathbf{x} + M\mathbf{x}') = \mathbf{e}^T M(\mathbf{x} + \mathbf{x}') = \mathbf{e}^T(\mathbf{x} + \mathbf{x}').$$

However, there are such fixed parity check row vectors that only have 12 nonzero bytes, and therefore involve only 12 active S-boxes. One such parity check row vector over GF(2) given in hexadecimal notation is

(00999900 CC5555CC 00999900 CC5555CC).

There are many further ways in which such parity checks can be used in the analysis of the AES. For example, we have seen above that the 126-dimensional subspace V'_{14} is M-invariant. Furthermore, the lower right 2×2 submatrix of $P^{-1}MP$ (Appendix B) shows that any coset of V'_{14} is mapped to itself by M. Thus we have identified a partition of either the state space or the set of differences into four subsets in which this partition is preserved by the augmented linear diffusion. These four cosets are defined by the two parity check row vectors over GF(2) given in hexadecimal notation by

(AAAAAAAA AAAAAAAA AAAAAAAA AAAAAAAA),
(5AF05AF0 5AF05AF0 5AF05AF0 5AF05AF0).

The potential of such observations [88] has not been explored to any great extent. In [89] some of this work was extended to similar properties over **F** rather than GF(2), and some high probability differential effects under related sequences of round keys were noted. While these observations do not apply to the AES, they demonstrate that high probability differential effects can be observed in AES-like ciphers satisfying the demands of the wide trail strategy [37, 39]. Thus further analysis of some of the issues raised in [87–89] may yet be of interest in the analysis of the AES.

2. Algebraic Representations

There can be many equivalent ways to describe a cryptosystem. Although standardisation requires the same convention to be used for data representation, alternative representations of the cipher operations can be of much interest. Some representations might be helpful to implementors, perhaps as a way of improving performance or providing additional protection against side-channel attacks. Other representations may be useful to the cryptanalyst in the hope that they provide further insights to the properties of the cipher.

Alternative representations of block ciphers are constructed by defining mappings. Suppose we have an original block cipher $\mathcal{E}$ with a state space $\mathcal{X}$ and key space $\mathcal{K}$, and a new block cipher $\mathcal{E}'$ with state space $\mathcal{X}'$ and key space $\mathcal{K}'$. We can now define a plaintext mapping σ, a ciphertext mapping γ, and a key mapping κ between the respective spaces of

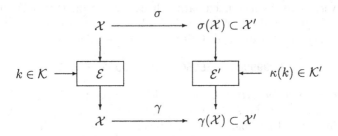

Figure 4.2. $\mathcal{E}'$: an alternative representation of the cipher $\mathcal{E}$.

the two block ciphers such that

$$\sigma, \gamma \colon \mathcal{X} \to \mathcal{X}' \text{ and } \kappa \colon \mathcal{K} \to \mathcal{K}'.$$

We say that the block cipher $\mathcal{E}'$ is an *alternative representation* of the block cipher $\mathcal{E}$ if, for all $x \in \mathcal{X}$ and $k \in \mathcal{K}$,

$$\mathcal{E}'_{\kappa(k)}(\sigma(x)) = \gamma(\mathcal{E}_k(x)).$$

An alternative representation is best illustrated by the commuting diagram of Figure 4.2. If the mapping functions are injective, then we can replicate encryption by $\mathcal{E}$ using the cipher $\mathcal{E}'$. We map the original plaintext to the new plaintext with σ and we map the original key to the new key with κ. We then encrypt the new plaintext with $\mathcal{E}'$ under the new key to obtain a new ciphertext. We can recover the original ciphertext from the new ciphertext. The recovered ciphertext is what would have been obtained if we had encrypted directly with the original block cipher $\mathcal{E}$. In this case, we say that the cipher $\mathcal{E}$ is *embedded* in the cipher $\mathcal{E}'$.

Cryptanalytic techniques for block ciphers can sometimes be described by using such commuting diagrams and their generalisations. For example, such a technique arises when a block cipher has *linear factors* or *linear structures* [16, 45, 106] and such properties have the potential to reduce the cost of key search by "factoring out" algebraically-related encryptions.

Alternative representations where the original and new block ciphers are identical ($\mathcal{X}' = \mathcal{X}$, $\mathcal{K}' = \mathcal{K}$ and $\mathcal{E}' = \mathcal{E}$) have been termed *self-dual* in [7]. The property of a cipher being self-dual under non-trivial affine mappings is essentially equivalent to the property of a cipher possessing linear factors or structures.

EXAMPLE 4.1 The complementation property of the DES gives a non-trivial self-dual cipher or, equivalently, a linear structure. Let $\mathbf{1}_x$ and $\mathbf{1}_k$ denote the vectors $(1, \ldots, 1)^T$ of lengths 64 and 56 respectively. Then take σ and γ to be the mapping $x \mapsto x + \mathbf{1}_x$, and κ to be the mapping $k \mapsto k + \mathbf{1}_k$. These are known as the complementation mappings and give a non-trivial self-dual cipher for the DES. Alternatively, we can obtain a linear structure by setting

$$\mathcal{X}' = \frac{\mathcal{X}}{\langle \mathbf{1}_x \rangle} \text{ and } \mathcal{K}' = \frac{\mathcal{K}}{\langle \mathbf{1}_k \rangle}$$

to be quotient spaces with respective natural mappings $\sigma = \gamma$ and κ. The equivalent commuting diagram to that of Figure 4.2 can be completed by setting

$$\mathcal{E}'_{\kappa(k)}(\sigma(x)) = \mathcal{E}_{k+\mathbf{1}_k}(x + \mathbf{1}_x) = \mathcal{E}_k(x) + \mathbf{1}_x. \qquad \square$$

EXAMPLE 4.2 The AES round function is self-dual. We define φ to be the permutation of the state or key array bytes defined by

$$\varphi = (\mathcal{S}_{00}\mathcal{S}_{03}\mathcal{S}_{02}\mathcal{S}_{01})(\mathcal{S}_{10}\mathcal{S}_{13}\mathcal{S}_{12}\mathcal{S}_{11})(\mathcal{S}_{20}\mathcal{S}_{23}\mathcal{S}_{22}\mathcal{S}_{21})(\mathcal{S}_{30}\mathcal{S}_{33}\mathcal{S}_{32}\mathcal{S}_{31}).$$

We define the three mappings ρ, γ, and κ to be the mappings of the state and key spaces induced by one of the permutations φ, φ^2, or φ^3 of the array of bytes. If we let $\mathcal{E}$ denote the AES round function, then

$$\mathcal{E}_{\kappa(k)}(\sigma(x)) = \gamma(\mathcal{E}_k(x)),$$

so the AES round function is self-dual [73]. However, this property does not extend to the full AES because of the action of the key schedule. $\square$

Such a framework for alternative representations can be extended stochastically. The standard statistical techniques of block cipher cryptanalysis, such as differential and linear cryptanalysis [10, 78], can be described using this simple generalisation. We simply view the commuting diagram as holding statistically and require that the function $\mathcal{E}'$ completes the commuting diagram with a suitable probability [86, 117].

Representations of the AES

A number of alternative representations have been proposed for the AES. They exploit the structure of the cipher and are mostly constructed by defining homomorphisms of the AES state and key spaces.

The state space of the AES is composed of 16 bytes, where each byte is considered as an element of the field **F**. The set $\mathbf{F}^{16}$ has both a vector space structure and a ring structure with component-wise multiplication.

We can therefore consider the state space of the AES with this natural algebraic structure as an **F**-algebra (Section 2.3), which we call the AES *state space algebra* [21].

The algebraic transformations of the state space algebra, that is transformations which preserve most of the structure of the algebra, are necessarily based on either a linear transformation or on a ring-theoretic transformation of the state space. As the AES round transformations are all algebraic operations, there are many opportunities to construct alternative representations. If an alternative representation is based on algebra isomorphisms, then we term the alternative representation an *isomorphic* cipher.

3.   Big Encryption System (BES)

One representation of the AES is derived by embedding the AES in a larger cipher called the *Big Encryption System (BES)* [89]. The BES is defined as a way to replicate the action of the AES using simple algebraic operations over **F**. The BES operates on 128-byte blocks with 128-byte keys and has a very simple algebraic structure. One round of the BES consists of inversion of each of these 128 bytes and an affine transformation with respect to a vector space of dimension 128 over **F**.

Embedding mapping of the AES in the BES

We denote the state space algebras of the AES by $\mathbf{A} = \mathbf{F}^{16}$ and of the BES by $\mathbf{B} = \mathbf{F}^{128}$. The mapping to embed the AES in the BES is based on the vector conjugate mapping $\phi : \mathbf{F} \to \mathbf{F}^8$, which maps an element of **F** to a vector of its eight conjugates. This mapping ϕ is an injective ring homomorphism given by

$$a \mapsto \left(a^{2^0}, a^{2^1}, a^{2^2}, a^{2^3}, a^{2^4}, a^{2^5}, a^{2^6}, a^{2^7} \right)^T .$$

This definition can be extended in the obvious way to an embedding function $\phi : \mathbf{A} \to \mathbf{B}$ defined by

$$(a_0, \ldots, a_{15})^T \mapsto (\phi(a_0), \ldots, \phi(a_{15}))^T ,$$

which is also an injective ring homomorphism. We can therefore use ϕ to embed an element of the AES state space **A** into the BES state space **B**, and we define

$$\mathbf{B_A} = \phi(\mathbf{A}) \subset \mathbf{B}$$

to be the embedded image of the AES state space. We note that $\mathbf{B_A}$ is a subring of **B** but not a subalgebra. However, $\mathbf{B_A}$ contains a basis for **B** as a vector space, and so **B** is the closure of $\mathbf{B_A}$ [21]. Since the

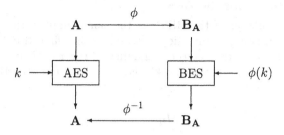

Figure 4.3. The relationship between the AES and the BES.

inverse mapping $\phi^{-1} : \mathbf{B_A} \to \mathbf{A}$ is well-defined, the BES gives rise to the commuting diagram of Figure 4.3.

Structure of the BES

A full description of the BES is given in [89]. The inversion, ShiftRows, MixColumns, and AddRoundKey operations in the AES with state space $\mathbf{A}$ are replaced by the obvious corresponding operations for the BES with state space $\mathbf{B}$. The addition of the S-box constant **63** in the AES is replaced by the obvious corresponding operation for the BES with state space $\mathbf{B}$, but then incorporated into a revised round key. The remaining operation yet unaccounted for is the GF(2)-linear mapping in the SubBytes S-box.

The GF(2)-linear mapping is defined by considering the Rijndael field $\mathbf{F}$ as a vector space of dimension 8 over GF(2) [95]. This is implicitly accomplished in the AES by the natural mapping $\psi \colon \mathbf{F} \to \mathrm{GF}(2)^8$. The componentwise AES GF(2)-linear operation $\mathbf{F} \to \mathbf{F}$ is then defined by $a \mapsto \psi^{-1}(\xi(\psi(a)))$, where $\xi : \mathrm{GF}(2)^8 \to \mathrm{GF}(2)^8$ is the linear transformation given in Section 3.2. It is the need for the maps ψ and ψ^{-1} which complicates the algebraic analysis of the AES. The GF(2)-linear mapping can be realised by a linearised polynomial over $\mathbf{F}$ (Section 2.4). The GF(2)-linear mapping $\mathbf{F} \to \mathbf{F}$ is therefore given by

$$a \mapsto \mathtt{05}a^{2^0} + \mathtt{09}a^{2^1} + \mathtt{F9}a^{2^2} + \mathtt{25}a^{2^3} + \mathtt{F4}a^{2^4} + \mathtt{01}a^{2^5} + \mathtt{B5}a^{2^6} + \mathtt{8F}a^{2^7}.$$

This means that the GF(2)-linear operation in the AES S-box can be defined within the BES by an 8×8 matrix over $\mathbf{F}$. This matrix replicates the AES action of the GF(2)-linear mapping on the first byte of a vector conjugate set and ensures that the property of vector conjugacy is preserved on the remaining bytes.

Round function of the BES

The round function of the BES has the same simple form as that of the AES, consisting of inversion followed by an affine transformation. Suppose that the state at the beginning of the round of the BES is $\mathbf{b} \in \mathbf{B}$ and that the BES round key is $(\mathbf{k}_B)_i \in \mathbf{B}$, then the BES round function is given by

$$\mathbf{b} \mapsto M_B \mathbf{b}^{(-1)} + (\mathbf{k}_B)_i,$$

where M_B is a 128×128 matrix over $\mathbf{F}$ performing linear diffusion within the BES [89]. The BES diffusion matrix M_B and the AES diffusion matrix M are closely related and share algebraic properties. M_B is sparse with minimal polynomial $(x+1)^{15}$ and related invariant subspaces. Furthermore, the round function of the AES can be expressed in terms of the BES. If the state at the start of an AES round is $\mathbf{w} \in \mathbf{A}$ and the round key is $\mathbf{k}_i^* \in \mathbf{A}$, then the AES round function is given by

$$\mathbf{w} \mapsto M\mathbf{w}^{(-1)} + \mathbf{k}_i^* = \phi^{-1} \left(M_B \left(\phi(\mathbf{w})^{(-1)} \right) + \phi(\mathbf{k}_i^*) \right).$$

Thus the AES can easily be defined in terms of the BES.

This definition of the AES round function in terms of the BES can allow certain algebraic properties of the AES to be seen directly. For example, the component output functions of the inversion operation are related by linear transformations, which gives various results about the AES S-box and AES round function [53, 120].

The AES embedding in the BES

The effect of the embedding mapping $\phi : \mathbf{A} \to \mathbf{B}$ on the AES encryption function is to induce an embedded encryption function $f_\phi : \mathbf{B}_A \to \mathbf{B}_A$. This function can be naturally extended to a function $\overline{f}_\phi : \mathbf{B} \to \mathbf{B}$ and so the BES can be naturally considered as the closure of the vector conjugate embedding of the AES [21].

Since the BES can be expressed using simple algebraic operations over a single field $\mathbf{F}$, this yields one particularly useful insight into the cipher. Using the BES we are able to obtain a multivariate quadratic equation system over $GF(2^8)$ that describes an AES encryption. As we see in Chapter 5, this system is sparser and simpler than the system obtained directly from the AES [89]. The generation and possible solution of such multivariate equation systems for the AES is the subject of Chapters 5 and 6.

4.   Other Representations of the AES

We now consider some other representations of the AES [7, 8, 105] which have been termed *dual ciphers*. We classify these representations by the properties of their representation mappings.

Isomorphic ciphers

Many of the dual ciphers [7, 8, 105] are alternative AES representations where the mappings of state and key spaces are algebra isomorphisms of the AES state space algebra. The resultant ciphers are therefore isomorphic to the AES.

The finite field $GF(2^8)$ can be constructed as an extension field of any of its subfields. Isomorphic representations of $GF(2^8)$ can thus be constructed from the chain of subfields

$$GF(2) \subset GF\left(2^2\right) \subset GF\left(2^4\right) \subset GF\left(2^8\right).$$

Each irreducible polynomial of degree d in $GF(2^n)[x]$ can be used to construct a finite extension of degree d of $GF(2^n)$ isomorphic to $GF(2^{nd})$ (Section 2.4).

EXAMPLE 4.3 The polynomial $x^4 + x + 1$ is irreducible in $GF(2)[x]$ and we denote one of its roots by ρ. Thus we have

$$\mathbb{K} = GF(2)(\rho) \cong \frac{GF(2)[x]}{\langle x^4 + x + 1 \rangle} \cong GF(2^4).$$

The polynomial $y^2 + y + \rho^3$ is irreducible in $\mathbb{K}[y]$. If we denote a root of $y^2 + y + \rho^3$ by ζ, then we have

$$\mathbb{K}(\zeta) = (GF(2)(\rho))\,(\zeta) \cong \frac{\mathbb{K}[y]}{\langle y^2 + y + \rho^3 \rangle} \cong GF(2^8).$$

We can then represent any element of $\mathbb{K}(\zeta)$ by $a_1\zeta + a_0$ for some $a_0, a_1 \in \mathbb{K}$. As a_0 and a_1 can be naturally represented by a hexadecimal character, we can represent such an element $a_1\zeta + a_0$ as the pair of hexadecimal characters (a_1, a_0). The element $\rho\zeta \in \mathbb{K}(\zeta)$ satisfies

$$(\rho\zeta)^8 + (\rho\zeta)^4 + (\rho\zeta)^3 + (\rho\zeta) + 1 = 0,$$

so there is a field isomorphism between the Rijndael field $\mathbf{F} = GF(2)(\theta)$, where θ is the Rijndael root, and $(GF(2)(\rho))\,(\zeta)$ given by $\theta \mapsto \rho\zeta$. In hexadecimal representation $02 \mapsto (2, 0)$. Another example is given by $05 \mapsto (4, 7)$ as

$$\theta^2 + 1 \mapsto (\rho\zeta)^2 + 1 = \rho^2(\zeta + \rho^3) + 1 = \rho^2\zeta + (\rho^5 + 1) = \rho^2\zeta + (\rho^2 + \rho + 1). \quad \square$$

		Subfield	
Degree	GF(2)	GF(2^2)	GF(2^4)
2	1	6	120
4	3	60	–
8	30	–	–

Figure 4.4. The number of irreducible polynomials over subfields of GF(2^8).

Different representations of GF(2^8) can be constructed using irreducible polynomials of the appropriate degree in the univariate polynomial rings over subfields of GF(2^8) [7, 8, 105]. The numbers of such irreducible polynomials are given in Figure 4.4. In total there are $30 + (1 \cdot 60) + (3 \cdot 120) + (1 \cdot 6 \cdot 120) = 1170$ different isomorphic representations of the Rijndael field **F** based on subfields. These different representations define 1170 ciphers isomorphic to the AES.

The Frobenius automorphism $z \mapsto z^2$ of **F** can be used to obtain further isomorphic ciphers. This field automorphism can be extended in the obvious way to give an algebra isomorphism $\tau : \mathbf{F}^{16} \to \mathbf{F}^{16}$ of the AES state space. For any function $f : \mathbf{F}^{16} \to \mathbf{F}^{16}$ used by the AES, we can define a function $f^{(2)} : \mathbf{F}^{16} \to \mathbf{F}^{16}$ by $x \mapsto \tau(f(\tau^{-1}(x)))$. We can then replace f by $f^{(2)}$ and the key k by $\tau(k)$ to obtain a new cipher $\mathcal{E}^{(2)}$. This new cipher is an alternative representation of the AES and satisfies

$$\mathcal{E}^{(2)}_{\tau(k)}(\tau(x)) = \tau(\mathcal{E}_k(x)).$$

This representation $\mathcal{E}^{(2)}$ is an isomorphic cipher and has been termed *square dual* [7, 8]. There are eight Frobenius automorphisms of **F**. If these Frobenius mappings are combined with subfield mappings, we can then construct 9360 ciphers isomorphic to the AES.

These alternative representations of the AES are based on field isomorphisms of **F**. Thus it seems unlikely that they are of cryptanalytic interest. However, such alternative representations have been proposed to improve the efficiency of hardware implementations, most particularly in the SubBytes transformation [39].

Regular representations

The regular representation is a standard and powerful mathematical technique for studying an algebra [23]. Regular representations of the AES state space algebra are discussed in [21].

A *representation* of an n-dimensional $\mathbb{F}$-algebra $\mathcal{A}$ is an algebra homomorphism from $\mathcal{A}$ to a subalgebra of the matrix algebra $\mathcal{M}_l(\mathbb{F})$ (Example 2.56). One standard representation is the *regular representation*.

This is the algebra homomorphism $\nu : \mathcal{A} \to \mathcal{M}_n(K)$ that maps $a \in \mathcal{A}$ to the matrix corresponding to the linear transformation $z \mapsto az$, where z is a vector over $\mathbb{F}$ of length n.

EXAMPLE 4.4 The complex numbers $\mathbb{C}$ form a 2-dimensional $\mathbb{R}$-algebra. The complex number $x + iy$ can be identified with its regular representation as a matrix, which is given by

$$\nu(x + iy) = \begin{pmatrix} x & y \\ -y & x \end{pmatrix}.$$

The set of all such matrices forms a 2-dimensional algebra over the real numbers and can be identified with the complex numbers $\mathbb{C}$. □

EXAMPLE 4.5 The Rijndael field $\mathbf{F}$ is an 8-dimensional GF(2)-algebra. The regular representation of $a \in \mathbf{F}$ is the 8×8 matrix T_a of Example 2.65. The set of all such matrices $\{T_a | a \in \mathbf{F}\}$ forms an 8-dimensional subalgebra of $\mathcal{M}_8(\text{GF}(2))$. The regular representation of the Rijndael field $\mathbf{F}$ as a GF(2)-algebra is this subalgebra. □

The AES state space can be considered as a GF(2)-algebra or as an $\mathbf{F}$-algebra. Example 4.5 shows that the regular representation of the AES state space as a GF(2)-algebra is a set of block diagonal matrices forming a subalgebra of $\mathcal{M}_{128}(\text{GF}(2))$. This regular representation is given by the mapping

$$\begin{pmatrix} a_0 \\ a_1 \\ \vdots \\ a_{15} \end{pmatrix} \mapsto \begin{pmatrix} T_{a_0} & 0 & \cdots & 0 \\ 0 & T_{a_1} & \cdots & 0 \\ \vdots & \vdots & \ddots & \vdots \\ 0 & 0 & \cdots & T_{a_{15}} \end{pmatrix},$$

where $(a_0, a_1, \ldots, a_{15})^T$ is interpreted as a vector over GF(2) of length 128. The AES encryption process can then be defined in terms of standard matrix operations.

- **Inversion.** For the block diagonal matrix A, this is the mapping $A \mapsto A^{(-1)} = A^{254}$. This is matrix inversion if A is invertible.

- **Augmented linear diffusion.** For the block diagonal matrix A, there exist block diagonal matrices D_i and permutation matrices P_i $(0 \le i \le 31)$ such that this linear transformation can be defined by

$$A \mapsto \sum_{i=0}^{31} D_i P_i A P_i^T.$$

- **Round key addition.** For the block diagonal matrix A and round key matrix K, this is the mapping $A \mapsto A + K$.

The AES state space is also an **F**-algebra, with regular representation given by the algebra homomorphism $\mathbf{F}^{16} \to \mathcal{M}_{16}(\mathbf{F})$ defined by

$$
\begin{pmatrix} x_0 \\ x_1 \\ \vdots \\ x_{15} \end{pmatrix} \mapsto \begin{pmatrix} x_0 & 0 & \cdots & 0 \\ 0 & x_1 & \cdots & 0 \\ \vdots & \vdots & \ddots & \vdots \\ 0 & 0 & \cdots & x_{15} \end{pmatrix}.
$$

Thus the regular representation of the AES state space $\mathbf{F}^{16}$ is $\mathcal{D}_{16}(\mathbf{F})$, the **F**-algebra of 16×16 diagonal matrices. Similarly, the matrix algebra $\mathcal{D}_{128}(\mathbf{F})$ of 128×128 diagonal matrices is the regular representation of the BES state space $\mathbf{F}^{128}$. This gives an embedding of an element of the AES subset of the BES defined by

$$
\begin{pmatrix} x_0^{2^0} \\ x_0^{2^1} \\ \vdots \\ x_0^{2^7} \\ x_1^{2^0} \\ \vdots \\ x_{15}^{2^7} \end{pmatrix} \mapsto \begin{pmatrix} x_0^{2^0} & 0 & \cdots & 0 & 0 & \cdots & 0 \\ 0 & x_0^{2^1} & \cdots & 0 & 0 & \cdots & 0 \\ \vdots & \vdots & \ddots & \vdots & \vdots & & \vdots \\ 0 & 0 & \cdots & x_0^{2^7} & 0 & \cdots & 0 \\ 0 & 0 & \cdots & 0 & x_1^{2^0} & \cdots & 0 \\ \vdots & \vdots & & \vdots & \vdots & \ddots & \vdots \\ 0 & 0 & \cdots & 0 & 0 & \cdots & x_{15}^{2^7} \end{pmatrix}.
$$

The regular representation of the AES subset of the BES is the subring of diagonal matrices where the octets form sets of conjugates. The BES encryption process, and hence the AES encryption process, can be defined in terms of the diagonal matrix B of a BES state space vector.

- **Inversion.** $B \mapsto B^{(-1)} = B^{254}$.

- **Linear Diffusion.** $B \mapsto \sum_{i=0}^{31} D_i P_i B P_i^T$, where D_i are diagonal matrices and P_i are permutation matrices.

- **Subkey Addition.** $B \mapsto B + K$, where K is the regular representation of the round key.

Logarithmic representations

The AES is specified using a polynomial representation for the elements of **F**. However we can also represent an element of **F** as an element of $\overline{\mathbb{Z}}_{255}$ by using the discrete logarithm (Section 2.4). We can thus give the *logarithm representation* of an element of the AES state space as an element of the set $\left(\overline{\mathbb{Z}}_{255}\right)^{16}$. Alternative representations based on the logarithmic representation of **F** are termed *log dual* ciphers in [7]. There are 128 primitive elements in **F** giving 128 different log dual ciphers to the AES. Full details of how to specify a logarithm representation of the AES are given in [7].

A multiplicative operation is easily formulated in the logarithm representation, and an additive operation can be defined in terms of the Zech logarithm (Section 2.4). We can directly represent an element of **F** as an element of $\overline{\mathbb{Z}}_{255}$ using the Zech logarithm and this gives a *Zech logarithm* representation of the AES state space as an element of the set $\left(\overline{\mathbb{Z}}_{255}\right)^{16}$. This could lead to a more succinct description of the AES than the conventional logarithm representation.

Identity reducible representations

The CES [82] is an alternative representation of the AES. However, representations such as the CES have a property that has been termed *identity reducible* and do not appear to provide any new perspectives on the AES [21].

5.   Group Theoretic Properties

Fundamentally, a block cipher provides a succinct description for an indexed set of permutations on the state space. Consequently we might try and gain insight into the structure of a block cipher by considering block cipher transformations as elements of some permutation group. Similarly, the constituent round functions of an iterated block cipher also form sets of permutations and may be analysed from the same perspective. The main theme of this section is to consider an analysis of the AES encryption and round function transformations as permutations acting on the AES state space.

Iterated block ciphers

Suppose a block cipher has a state space $\mathcal{X}$ and a key space $\mathcal{K}$. For a given key $k \in \mathcal{K}$, encryption under the block cipher is a permutation $\varepsilon_k : \mathcal{X} \rightarrow \mathcal{X}$. The set $\mathcal{E} = \{\varepsilon_k | k \in \mathcal{K}\}$ of all possible encryptions of the block cipher is a subset of $S_{\mathcal{X}}$, the group of all permutations of the state space. The group $\mathcal{G} = \langle \varepsilon_k | k \in \mathcal{K} \rangle$ generated by the set of cipher permutations is known as the *group generated by the cipher*. If $\mathcal{G} = \mathcal{E}$, that is the set of permutations $\{\varepsilon_k | k \in \mathcal{K}\}$ forms a group, then we say that the cipher is a group. As $\mathcal{G}$ is a finite group, the cipher is a group if and only if the set $\mathcal{E}$ is closed under the operation of composition. For such a cipher, multiple encryption offers no extra security over single encryption.

More generally, certain properties of the group $\mathcal{G}$ generated by a cipher are of interest cryptographically [60] and attacks have been proposed against ciphers that do not satisfy some of these properties [65, 101]. However good group theoretic properties are not sufficient to guarantee a strong cipher [85].

Computing the group $\mathcal{G}$ generated by a block cipher is often difficult. Let υ_{k_i} denote the round function of the cipher under the subkey $k_i \in \mathcal{K}_s$, where $\mathcal{K}_s$ is the space of round subkeys. The round functions υ_{k_i} that make up an encryption ε_k are also permutations of the state space $\mathcal{X}$, and it is often easier to calculate the various groups generated by these permutations. Suppose we have an r-round block cipher with a key schedule function $KS : \mathcal{K} \to (\mathcal{K}_s)^r$, so that any key $k \in \mathcal{K}$ gives rise to r subkeys in $\mathcal{K}_s$. The round function permutations naturally suggest the following three groups of relevance to the block cipher:

$$
\begin{aligned}
\mathcal{R} &= \langle \upsilon_k \mid k \in \mathcal{K}_s \rangle, \\
\mathcal{P} &= \langle \upsilon_{k_r} \dots \upsilon_{k_2} \upsilon_{k_1} \mid k_i \in \mathcal{K}_s \rangle, \\
\mathcal{G} &= \langle \upsilon_{k_r} \dots \upsilon_{k_2} \upsilon_{k_1} \mid KS(k) = (k_1, \dots, k_r) \rangle = \langle \varepsilon_k \mid k \in \mathcal{K} \rangle.
\end{aligned}
$$

Thus $\mathcal{R}$ is the group generated by the round functions and $\mathcal{P}$ is the group generated by an arbitrary composition of r round functions. The group $\mathcal{G}$ generated by the cipher can also be regarded as the group generated by any composition of r round functions permitted by the key schedule. The relationship between these groups is that $\mathcal{G}$ is a subgroup of $\mathcal{P}$, and $\mathcal{P}$ is a normal subgroup of $\mathcal{R}$ ($\mathcal{G} < \mathcal{P} \lhd \mathcal{R}$). Thus the group generated by the round functions upper bounds the group generated by the cipher.

EXAMPLE 4.6 Properties of the groups generated by the DES have been extensively researched. Initially, it was observed that the cycle structures of certain permutations could be used to provide a lower bound on the order of the group $\mathcal{G}$ generated by DES [24]. Subsequently, the cycle structures were extensively analysed [15, 65, 83, 103, 104] and were used to show that $|\mathcal{G}| > 2^{56}$, so the DES is not a group [15, 25].

The DES round function under any key, and hence any DES encryption, is an even permutation. Furthermore, the group $\mathcal{R}$ generated by the round functions of the DES is the alternating group $A_{2^{64}}$ on the state space of the DES, which is a large, simple, primitive and highly transitive group [118]. Thus the group $\mathcal{P} \lhd \mathcal{R}$ generated by the composition of any fixed number round functions is also the alternating group $A_{2^{64}}$. It follows that the group $\mathcal{G}$ generated by the DES is a subgroup of the alternating group $A_{2^{64}}$, although little more is known about its structure.   □

Cycle structures

We now discuss the cycle structures of the different operations in the AES round function when considered as permutations of the state space. Some related results are given in [40, 73].

We first consider the permutation π_{ij} of the AES state space of size 2^{128} given by the application of the AES S-box to the byte $\mathcal{S}_{ij}$, whilst fixing the other fifteen bytes. The action of the SubBytes operation on the AES state space is given by the permutation $\pi_{00} \ldots \pi_{33}$. This permutation is the 16-fold product of permutations with the same cycle structure. Thus the SubBytes operation is an even permutation of the AES state space.

We now consider the ShiftRows and MixColumns operations. We saw in Section 4.1 that these operations could be defined in terms of the application of the permutation matrix $\widehat{R}$ to four bytes of the AES state space, whilst fixing the other 12 bytes, with respect to some basis. We denote the permutation of the AES state space given by an application of the permutation matrix $\widehat{R}$, with respect to the appropriate basis, to a row i by $\pi^{(i\cdot)}$ and to a column j by $\pi^{(\cdot j)}$. The ShiftRows operation and MixColumns operations are then given, respectively, by the permutations

$$\left(\pi^{(1\cdot)}\right)\left(\pi^{(2\cdot)}\right)^2\left(\pi^{(3\cdot)}\right)^3 \text{ and } \left(\pi^{(\cdot 0)}\right)\left(\pi^{(\cdot 1)}\right)\left(\pi^{(\cdot 2)}\right)\left(\pi^{(\cdot 3)}\right).$$

These are 6-fold and 4-fold products of permutations with the same cycle structure. Thus both the ShiftRows operation and the MixColumns operation are even permutations of the AES state space.

We finally consider the AddRoundKeys operation. It is clear that the AddRoundKeys operation is the product of 2^{127} transpositions for a nonzero round key and the identity transformation for a zero round key. Thus the AddRoundKeys operation is an even permutation.

All operations used by the AES round functions are therefore even permutations of the AES state space, and so we have Theorem 4.7.

THEOREM 4.7 *The AES round function is an even permutation.*

The cycle structures of each component of the AES round function can be largely deduced from Figure 4.1 and are given in Figure 4.5. We note that the AES S-box permutation on the 2^8 elements of $\mathbf{F}$ has five disjoint cycles of lengths 87, 81, 59, 27 and 2, and so is an odd permutation [119]. Detailed analysis of various permutations generated by the AES round functions gives Theorem 4.8 [119].

THEOREM 4.8 *The group* $\mathcal{R}$ *generated by the AES round functions is the alternating group* $A_{2^{128}}$.

| | | Number of cycles | | Permutation |
	fixed	order 2	order 4	parity
Inversion	2^{16}	$2^{-1}(2^{128} - 2^{16})$	0	Even
GF(2)-linear	2^{48}	$2^{-1}(2^{96} - 2^{48})$	$2^{-2}(2^{128} - 2^{96})$	Even
S-box constant	0	2^{127}	0	Even
ShiftRows	2^{64}	$2^{-1}(2^{96} - 2^{64})$	$2^{-2}(2^{128} - 2^{96})$	Even
MixColumns	2^{32}	$2^{-1}(2^{64} - 2^{32})$	$2^{-2}(2^{128} - 2^{64})$	Even
AddRoundKey	0	2^{127}	0	Even
(0 Round Key)	(2^{128})	(0)	(0)	(Even)

Figure 4.5.   The cycle structure of the different AES components.

This means that the group $\mathcal{P}$ for the AES is also the alternating group. This implies that "from the algebraic point of view some thinkable weakness [of the AES] can be excluded" [119]. We note that the group $\mathcal{G}$ generated by the AES is not just a simple composition of round functions, since there is an initial AddRoundKey operation and no MixColumns operation in the final round. However the resulting round operations are even permutations, and so we also have $\mathcal{G} < A_{2^{128}}$.

Byte diffusion group

Suppose that G is a group of permutations of the 16 state array bytes, so $G < S_{16}$. Each element $g \in G$ can be used to define a permutation $g_{\mathbf{A}} \in S_{\mathbf{A}}$ of the AES state space $\mathbf{A} = \mathbf{F}^{16}$, which we define in Figure 4.6. Furthermore, such mappings can be extended to give the mappings $(gg')_{\mathbf{A}}$, $g_{\mathbf{A}} + g'_{\mathbf{A}}$, and $\lambda \cdot g_{\mathbf{A}}$ $(g, g' \in G$ and $\lambda \in \mathbf{F})$ of the AES state space $\mathbf{A}$. These mappings are also defined in Figure 4.6. Thus any formal sum

$$\sum_{g \in G} \lambda_g g \qquad [\lambda_g \in \mathbf{F}]$$

can be used to define a mapping $\left(\sum_{g \in G} \lambda_g g \right)_{\mathbf{A}}$ of the AES state space $\mathbf{A}$ given by

$$(\mathcal{S}_{00}, \ldots, \mathcal{S}_{33})^T \mapsto \sum_{g \in G} \lambda_g \left(\mathcal{S}_{g(00)}, \ldots, \mathcal{S}_{g(33)} \right)^T.$$

The set of all such formal sums of elements of G under the obvious rules of addition and multiplication forms an algebra, known as the *group algebra* [23] of G over $\mathbf{F}$, and is denoted by $\mathbf{F}[G]$.

We now discuss the permutations of the 16 state array bytes needed to define the ShiftRows operation and the MixColumns operation. Both

Mapping	Definition
$g_{\mathbf{A}}$	$(S_{00}, \ldots, S_{33})^T \mapsto \left(S_{g(00)}, \ldots, S_{g(33)}\right)^T$
$(gg')_{\mathbf{A}}$	$(S_{00}, \ldots, S_{33})^T \mapsto \left(S_{g(g'(00))}, \ldots, S_{g(g'(33))}\right)^T$
$g_{\mathbf{A}} + g'_{\mathbf{A}}$	$(S_{00}, \ldots, S_{33})^T \mapsto \left(S_{g(00)} + S_{g'(00)}, \ldots, S_{g(33)} + S_{g'(33)}\right)^T$
$\lambda \cdot g_{\mathbf{A}}$	$(S_{00}, \ldots, S_{33})^T \mapsto \left(\lambda S_{g(00)}, \ldots, \lambda S_{g(33)}\right)^T$

Figure 4.6. Definitions of some mappings on the AES state space **A**.

of these operations can be defined in terms of a specific permutation of the 16 bytes of the state array. Thus each operation can be defined in terms of a single element of the group algebra $\mathbf{F}[S_{16}]$. The permutation

$$\varrho = (S_{10}S_{13}S_{12}S_{11})(S_{20}S_{22})(S_{21}S_{23})(S_{30}S_{31}S_{32}S_{33})$$

is the byte permutation defined by the ShiftRows operation. Thus the group algebra element $\Delta = 1 \cdot \varrho \in \mathbf{F}[S_{16}]$ gives the mapping $\Delta_{\mathbf{A}}$, which is the ShiftRows operation. Similarly, a simultaneous rotation of all the columns in the state array by one position defines the permutation

$$\varsigma = (S_{00}S_{10}S_{20}S_{30})(S_{01}S_{11}S_{21}S_{31})(S_{02}S_{12}S_{22}S_{32})(S_{03}S_{13}S_{23}S_{33}).$$

We can then define the group algebra element $\Gamma \in \mathbf{F}[S_{16}]$ by

$$\Gamma = \theta \cdot e + (\theta + 1) \cdot \varsigma + 1 \cdot \varsigma^2 + 1 \cdot \varsigma^3,$$

where e denotes the identity element of S_{16}. The MixColumns operation is then given by the mapping $\Gamma_{\mathbf{A}}$. The byte diffusion of the round function of the AES is given by the ShiftRows operation followed by the MixColumns operation. Thus $\Lambda_{\mathbf{A}} = (\Gamma\Delta)_{\mathbf{A}}$ specifies the diffusion, where the group algebra element $\Lambda \in \mathbf{F}[S_{16}]$ is given by

$$\Lambda = \Gamma\Delta = \theta \cdot \varrho + (\theta + 1) \cdot \varsigma\varrho + 1 \cdot \varsigma^2\varrho + 1 \cdot \varsigma^3\varrho.$$

The mixing between the bytes required by the wide trail strategy is therefore given by the two elements Δ and Γ of the group algebra $\mathbf{F}[S_{16}]$, which depend entirely on the two permutations ϱ and ς of S_{16}. Thus if we define the subgroup $H = \langle \varrho, \varsigma \rangle < S_{16}$, then the ShiftRows operation and the MixColumns operation are given by elements of the smaller group algebra $\mathbf{F}[H]$. The byte diffusion within the AES required by the wide trail strategy can therefore be defined in terms of the group algebra $\mathbf{F}[H]$, and so we term H the *byte diffusion group* of the AES.

We now consider this byte diffusion group $H = \langle \varrho, \varsigma \rangle$. We first note that both ϱ and ς are even permutations so $H < A_{16}$. In order to discuss the diffusion group H, we define the permutation $\varphi = \varsigma\varrho\varsigma^{-1}\varrho^{-1} \in H$, so

$$\varphi = (S_{00}S_{03}S_{02}S_{01})(S_{10}S_{13}S_{12}S_{11})(S_{20}S_{23}S_{22}S_{21})(S_{30}S_{33}S_{32}S_{31}).$$

The permutation φ corresponds to a simultaneous rotation of all rows by one position. We further note that φ commutes with both ς and ϱ, so $\Lambda(1 \cdot \varphi) = (1 \cdot \varphi)\Lambda$. This observation shows that φ is a permutation of state array bytes that can be used to define a self-dual round function for the AES (Example 4.2).

In order to describe the byte diffusion group H, we define the subgroup $H_0 = \langle \varsigma \rangle < H$ and the normal subgroup $N = \langle \varrho, \varphi \rangle \lhd H$. Elements of the subgroup H_0 permute the rows of the state array, whereas elements of the normal subgroup N act entirely within each row. The subgroup H_0 is isomorphic to the cyclic group C_4 with four elements, and the normal subgroup N is isomorphic to $C_4 \times C_4$ and so has 16 elements. Furthermore, we can show that any element of H can be expressed uniquely as a product of an element of its normal subgroup N and an element of its subgroup H_0. It follows that the byte diffusion group H of the AES is the *semidirect product* [23] of N by H_0 and so has order 64.

This formulation of the byte diffusion group H shows that its action on the state array has two distinct parts. It can be divided into the action of the normal subgroup N acting entirely within each row and the action of the subgroup H_0 which permutes the rows. Such a property means that H is an imprimitive group (Section 2.1) and that each row of the state array is a block of imprimitivity. More generally, we have shown that any byte diffusion generated by the linear diffusion (ShiftRows and MixColumns) part of the AES round function can be given by a formal sum of elements from a small byte diffusion group H of size 64. This AES byte diffusion group H is small and structured in comparison to a possible byte diffusion group S_{16}.

Such analysis extends to the BES state space $\mathbf{F}^{128}$. A byte diffusion group for the BES of size 512 is obtained, which is a subgroup of A_{128}. This diffusion group is isomorphic to the direct product (Example 2.5) of the cyclic group with 8 elements and the AES diffusion group H.

Geometric properties

The operations used in the AES can be viewed as geometrical transformations. Inversion in a finite field is a geometrical transformation in projective geometry, while the augmented diffusion and round key addition form an affine transformation of a vector space. This leads to some observations about some geometric properties of the AES. These observations are projective in nature and are discussed more fully in [61]. We briefly discussed projective spaces in Section 2.5. However, our discussion below is based on the projective line $\overline{\mathbf{F}}$ of the Rijndael field $\mathbf{F}$, which we define in Definition 4.9.

DEFINITION 4.9 Let **F** be the Rijndael field and $\mathbf{F}^2$ the vector space of dimension 2 over **F**. The *projective line* $\overline{\mathbf{F}}$ of the Rijndael field is the set of one-dimensional subspaces of the vector space $\mathbf{F}^2$ under the action of the group of invertible linear transformations of $\mathbf{F}^2$.

The points of the projective line $\overline{\mathbf{F}}$ are all the one-dimensional subspaces of $\mathbf{F}^2$, so

$$\overline{\mathbf{F}} = \{ \langle (1, z) \rangle \mid z \in \mathbf{F} \} \cup \{ \langle (0, 1) \rangle \}.$$

The projective point $\langle (0, 1) \rangle$ is known as the *point at infinity*, so we can regard the projective line $\overline{\mathbf{F}}$ as $\mathbf{F} \cup \{\infty\}$. The group of transformations on $\overline{\mathbf{F}}$ is known as the *Projective General Linear Group* PGL(2, **F**), a group of order $2^8(2^{16} - 1)$. Furthermore, PGL(2, **F**) is a sharply triply transitive group (Section 2.1), which means that action of an element on three projective points uniquely identifies that element.

The potential for a geometrical approach can be seen by the analysis of two simple block ciphers with state spaces **F** and $\overline{\mathbf{F}}$ respectively given in Example 4.10.

EXAMPLE 4.10 Two ciphers $\mathcal{C}$ and $\overline{\mathcal{C}}$ with state spaces **F** and $\overline{\mathbf{F}}$ respectively are defined below (with the conventional interpretation for ∞).

Cipher	State Space	Round Key	Round Function	Definition
$\mathcal{C}$	**F**	$k \in \mathbf{F}$	$f_k : \mathbf{F} \to \mathbf{F}$	$x \mapsto x^{(-1)} + k$
$\overline{\mathcal{C}}$	$\overline{\mathbf{F}}$	$k \in \mathbf{F}$	$\overline{f_k} : \overline{\mathbf{F}} \to \overline{\mathbf{F}}$	$x \mapsto \frac{1}{x} + k$

The round functions f_k and $\overline{f_k}$ of the two ciphers agree on **F** unless a 0-inversion takes place. Thus the ciphers $\mathcal{C}$ and $\overline{\mathcal{C}}$ transform a given plaintext to the same ciphertext for most plaintexts.

We now consider the two groups

$$\begin{aligned} \mathcal{R} &= \langle f_k \mid k \in \mathbf{F} \rangle \cong \mathrm{Sym}(\mathbf{F}) \\ \text{and} \quad \overline{\mathcal{R}} &= \langle \overline{f_k} \mid k \in \mathbf{F} \rangle \cong \mathrm{PGL}(2, \mathbf{F}) \end{aligned}$$

generated by the round functions for the two ciphers (Section 4.5). We see that the round functions of $\mathcal{C}$ generate the symmetric group on **F**, so we would require many plaintext-ciphertext pairs to determine the cipher transformation. By contrast, the round functions of $\overline{\mathcal{C}}$ generate PGL(2, **F**), so only three plaintext-ciphertext pairs are needed to determine the cipher transformation.

However, $\mathcal{C}$ and $\overline{\mathcal{C}}$ encrypt most plaintexts in the same way. Thus the overall cipher transformation of $\mathcal{C}$ can also be determined with three plaintext-ciphertext pairs with high probability. $\qquad\square$

The analysis given in Example 4.10 is essentially a group-theoretic explanation of the interpolation attack [62, 63] for this type of block cipher. Furthermore, Example 4.10 shows that a practically insignificant change to the definition of the group action being considered can yield a very different group. Moreover, in this case the group $\overline{\mathcal{R}}$ generated by the modified group action is a far more accurate indicator of the cipher's resistance to an algebraic attack than the group $\mathcal{R}$ generated by the unmodified group action.

Further discussion of the projective aspects of the AES is given in [1, 27, 61]. Such analysis yields results such as the characterisation of the difference table used in differential cryptanalysis [10, 11] for the AES inversion operation [61]. Similar projective constructions for the entire AES state space $\mathbf{F}^{16}$ may yet be of future interest.

Chapter 5

EQUATION SYSTEMS FOR THE AES

The idea of breaking a cryptosystem by solving a system of equations is not new. Shannon states in his landmark paper that breaking a cryptosystem should require:

> ...as much work as solving a system of simultaneous equations in a large number of variables [113].

Even prior to this, the cryptanalysis of many historical ciphers might be described within such a framework. For example, some devices used to analyse the Enigma cipher at Bletchley Park, such as the *bombe*, were fundamentally devices that checked the consistency of equation systems [64].

More recently, some early attempts to understand and cryptanalyse the DES were based on describing a DES encryption as an equation system. An early report on the DES [56] explicitly considered the task of writing bits of the output from the DES S-boxes as equations involving the six input bits to an S-box and using such expressions as a basis for an attack. However this report was optimistic in stating that:

> "These expressions indicate that an attempt to solve for K [key] in terms of P [plaintext] and C [ciphertext] may result in a simpler set of equations than one would expect."

Following a more detailed analysis of the DES equation system [109], such attempts to analyse the DES by solving an equation system were largely abandoned. However, as motivation for the analysis of an AES equation system, we observe that equations lying at the heart of the AES appear to be much simpler than those for the DES. We can illustrate this by considering the expression for the most significant output bit x_1 of the first DES S-box in terms of the six input bits $w_1w_2w_3w_4w_5w_6$ [109].

This expression is given by

$$
\begin{aligned}
x_1 \;=\;\; & w_1w_2w_3w_4w_6 + w_1w_2w_3w_5w_6 + w_1w_2w_3w_4 + w_1w_2w_4w_5 \\
+\;\; & w_1w_2w_4w_6 + w_1w_2w_5w_6 + w_1w_3w_4w_5 + w_1w_3w_4w_6 \\
+\;\; & w_1w_2w_4 + w_1w_3w_4 + w_1w_3w_5 + w_1w_4w_6 + w_2w_3w_4 \\
+\;\; & w_3w_4w_5 + w_3w_4w_6 + w_4w_5w_6 \\
+\;\; & w_1w_4 + w_1w_5 + w_2w_3 + w_3w_4 \\
+\;\; & w_1 + w_2 + w_3 + w_5 + w_6 + 1.
\end{aligned}
$$

Even if we consider equations between the input bits and all of the output bits $x_1x_2x_3x_4$ of the first DES S-box, the simplest equation we obtain is the quadratic equation

$$
\begin{aligned}
0 \;=\;\; & w_1w_2 + w_2w_3 + w_2w_4 + w_2w_5 + w_2w_6 + w_1x_1 + w_2x_1 + w_3x_1 \\
+\;\; & w_4x_1 + w_5x_1 + w_6x_1 + w_1x_2 + w_2x_2 + w_3x_2 + w_4x_2 + w_5x_2 \\
+\;\; & w_6x_2 + w_1x_3 + w_2x_3 + w_3x_3 + w_4x_3 + w_5x_3 + w_6x_3 + w_1x_4 \\
+\;\; & w_3x_4 + w_4x_4 + w_5x_4 + w_6x_4 + x_1x_4 + x_2x_4 + x_3x_4 \\
+\;\; & w_1 + w_2 + w_3 + w_4 + w_5 + w_6 + x_1 + x_2 + x_3 + x_4 + 1.
\end{aligned}
$$

This complicated equation is the only quadratic equation relating the input and output bits of the first DES S-box. By contrast, there is a very simple quadratic equation over the Rijndael field $\mathbf{F}$ that relates the input and output of the AES inversion, which is the only operation in the AES that is not linear over $\mathrm{GF}(2)$. The relation over $\mathbf{F}$ between the non–zero input w and the output x of inversion is given by

$$
wx = 1.
$$

This simple fact has spurred much of the research we discuss in the following chapters.

1.   Basic Approaches

The simplest equations, at least in terms of the number of variables, would be equations in the plaintext, ciphertext, and key, as described in the above quotation from [56]. In this section, we describe one generic method and one method specific to the AES for deriving such equations for a block cipher encryption.

Interpolation

The *interpolation attack* is a method for the cryptanalysis of a block cipher whose encryption function can be expressed in terms of a univariate polynomial function of moderate degree [63]. Suppose we have such a block cipher with state space given by a finite field $\mathbb{F}$ and $(d+1)$

plaintext–ciphertext pairs $(p_i, c_i) \in \mathbb{F}^2$ $(i = 0, \ldots, d)$. The Lagrange Interpolation Formula (Theorem 2.28) states that the unique polynomial function $f : \mathbb{F} \to \mathbb{F}$ mapping p_i to c_i is given by

$$f(x) = \sum_{i=0}^{d} c_i \prod_{\substack{j=0 \\ j \neq i}}^{d} \left(\frac{x - p_j}{p_i - p_j} \right).$$

If the block cipher encryption can be expressed as a polynomial function of moderate degree d, then the encryption operation is given by the above polynomial function f. This function can then be used to encrypt any plaintext, or to decrypt any ciphertext, without knowledge of the secret key. The attack can be slightly modified to give an equation containing the key, which would then allow recovery of the secret key.

The interpolation attack can be adapted to some block ciphers that use the same inversion mapping $w \mapsto w^{(-1)}$ as used in the AES. Thus the interpolation attack illustrates some of the potential issues involved in using simple algebraic operations within an iterative cipher, even if these components can be used to make a block cipher that is extremely resistant to other types of cryptanalysis.

Algebraic expressions

Some algebraic expressions for an AES encryption were given in [52]. We now discuss how to derive such an expression.

The AES S-box consists of the composition of three simple algebraic operations, namely an inversion operation, the GF(2)-linear mapping and the addition of the S-box constant 63 (Section 3.2). The inversion operation in the S-box is given by $w \mapsto w^{(-1)} = w^{254}$, and we saw in Section 4.1 that the GF(2)-linear mapping in the S-box is given by the linearised polynomial function

$$x \mapsto 05x^{2^0} + 09x^{2^1} + \text{F}9x^{2^2} + 25x^{2^3} + \text{F}4x^{2^4} + 01x^{2^5} + \text{B}5x^{2^6} + 8\text{F}x^{2^7}.$$

The final part of the S-box is the addition of the S-box constant 63. Thus the AES S-box is given by the polynomial function over $\mathbf{F}$

$$w \mapsto \left(\sum_{i=0}^{7} \lambda_i w^{255 - 2^i} \right) + 63,$$

where the coefficients λ_i are given above.

This form for the modified AES S-box can be used to express the full AES encryption operation as a set of 16 polynomials over $\mathbf{F}$. Each polynomial expresses a particular ciphertext byte as a function of the plaintext and key bytes. However, these polynomials are extremely large and dense, and they are not practically soluble.

We can simplify these expressions under the assumption that no 0-inversion occurs, an event that happens with probability 0.53 (Section 3.2). We can then incorporate the S-box constant **63** as part of a modified key schedule to give a modified S-box consisting of just the inversion operation and the GF(2)-linear mapping (Section 4.1). For a nonzero input w, the modified S-box is given by the mapping

$$w \mapsto \sum_{i=0}^{7} \lambda_i w^{-2^i}.$$

We can then use this expression for the modified S-box to give an expression for an AES encryption using a form of continued fractions. In this way, each byte of the state space after five AES rounds, $\mathcal{S}_{i,j}^{(5)}$, is given by

$$\mathcal{S}_{i,j}^{(5)} = K + \sum \cfrac{C_5}{K^* + \sum \cfrac{C_4}{K^* + \sum \cfrac{C_3}{K^* + \sum \cfrac{C_2}{K^* + \sum \cfrac{C_1}{K^* + p_*^*}}}}} \, ,$$

where p, C_i and K correspond to plaintext bytes, known constants, and expanded key bytes respectively, with * indicating a known exponent or subscript [52].

A fully expanded expression for five rounds has around 2^{25} terms of the type $\frac{C}{K^* + p_*^*}$, whilst the expression for the full 10-round AES encryption has around 2^{50} terms. A type of *meet-in-the-middle* approach, using the expressions for encryption for the first five rounds and decryption for the remaining five rounds, has been proposed to obtain equations with around 2^{26} terms [52].

This technique gives rise to very compact algebraic expressions for an AES encryption operation. Other block ciphers do not seem to have this property. For example, it is estimated that an algebraic expression for a DES encryption would contain 2^{64} terms [52]. The AES, which has a much larger key and block size, nevertheless has a much smaller algebraic expression for encryption. While there is no algorithm that can practically solve these types of equations, their existence provides additional motivation in the search for more amenable systems of equations describing an AES encryption.

2.   Equation Systems over GF(2)

An AES equation system consists of two parts, namely an equation system for encryption and an equation system for the key schedule. The equation system for one encryption treats the plaintext and ciphertext values as constants and uses state variables specific to that encryption. Thus different encryptions have different equation systems. The equation system for the key schedule depends only on key variables, and therefore is common to all encryptions under the same key.

This section gives a derivation for an equation system over $GF(2)$ that describes an AES encryption. The AES key schedule uses the same operations as the AES encryption function, so the derivation of a corresponding equation system is very similar. This equation system uses state variables given by the input and output of the inversion mapping at every round and, as key variables, the round keys. Thus this basic equation system for an AES encryption has $10 \cdot 128 = 1280$ inversion input and output variables, giving a total of 2560 state variables. The equation system for an AES encryption can be expressed in terms of these 2560 state variables and $11 \cdot 128 = 1408$ round key variables.

Linear equations

In Section 4.1, we show that the augmented diffusion matrix M over $GF(2)$ can be used to map a vector representing the output of all 16 AES inversions in one round to a vector representing the input to the AES inversions in the succeeding round. This mapping is given by $\mathbf{x} \mapsto M\mathbf{x} + \mathbf{k}_i^*$, where $\mathbf{x}$ is this inversion output vector and $\mathbf{k}_i^*$ is the modified round key. There are similar affine mappings which relate the plaintext to the input of the AES inversions in the first round and the output of the AES inversions in the final round to the ciphertext. Thus there are $11 \cdot 128 = 1408$ linear equations in this system. Furthermore, these linear equations are very sparse due to the round structure of the AES.

Nonlinear equations

The nonlinear relations in the AES equation system arise from the AES inversion operation. In order to give a complete equation system over $GF(2)$ for encryption, we need to establish nonlinear equations relating the components of the input and output of an AES inversion function.

The use of multivariate quadratic equations over $GF(2)$ to describe the relationship between inversion input and output was discussed in [31, 32]. We give a simple derivation based on linear algebra. We note however that the multivariate equations presented here differ from the ones given in Appendix A of [31], which consisted of multivariate quadratic equations for the whole S-box operation rather than just the inversion operation. However, as the output of SubBytes is the result of an affine mapping over $GF(2)$ of the inversion output, a simple linear substitution of the x variables in the equations of Appendix A gives those in [31].

The defining relation over $\mathbf{F}$ between the input w and the output x of an AES inversion is clearly $wx = 1$ (unless both are zero). We now consider how this relationship between the field elements w and x can be translated to their components when w and x are considered as vectors of length 8 over $\mathrm{GF}(2)$.

The mapping $z \mapsto \theta z$ describing the multiplication by θ in $\mathbf{F}$ is given in the vector space $\mathrm{GF}(2)^8$ by the linear transformation $z \mapsto T_\theta z$, where the matrix T_θ is given in Example 2.65. This can be extended to the mapping $z \mapsto wz$ describing the multiplication by any element $w \in \mathbf{F}$. This multiplication corresponds to the linear transformation $z \mapsto C_w z$, where C_w is a 8×8 matrix over $\mathrm{GF}(2)$ with the vector $T_\theta^{8-i} w$ in column i, that is

$$C_w = \left(\, T_\theta^7 w \mid T_\theta^6 w \mid \ldots \mid T_\theta w \mid w \,\right).$$

The vector in $\mathrm{GF}(2)^8$ corresponding to the product wx in the field $\mathbf{F}$ is given by $C_w x$ (Appendix A).

For the case of the AES inversion, we have

$$C_w x = (0,0,0,0,0,0,0,1)^T$$

unless a 0-inversion has taken place. In this case, we have $w = x = 0$, and $C_w x$ is the zero vector. Thus the first seven components of $C_w x$ are identically zero, which give us seven quadratic equations over $\mathrm{GF}(2)$ in the components of w and x. Unless a 0-inversion takes place, the last component is identically 1, giving us a further equation over $\mathrm{GF}(2)$ with probability $\frac{255}{256}$. Thus the above matrix equation gives seven multivariate quadratic equations over $\mathrm{GF}(2)$ as well as another equation with high probability. Furthermore, we note that these are bilinear equations in the w and x variables.

In addition to the equations above, we can derive further equations from the field equation $wx = 1$. We clearly have $wx^2 = x$ and $w^2 x = w$. These two equations can be expressed as the matrix equations

$$(C_w S + I)x = 0 \text{ and } (C_x S + I)w = 0,$$

where S is the matrix for the squaring map (Example 2.65). These vectors are also given in Appendix A. These two matrix equations give us 16 further multivariate quadratic equations over $\mathrm{GF}(2)$. These are biaffine equations in the w and x variables.

We note that the equations $wx^4 = xx^2$ and $w^4 x = ww^2$, which are equivalent to the matrix equations

$$(C_w S^2 + C_x)x = 0 \text{ and } (C_x S^2 + C_w)w = 0,$$

also give rise to 16 further quadratic equations over $\mathrm{GF}(2)$.

In total we have 39 multivariate quadratic equations over GF(2) that relate the input and output of an AES inversion. In addition, a further equation is valid with high probability. However it may sometimes be advantageous just to concentrate on the simpler bilinear or biaffine equations, obtaining a total of 23 multivariate quadratic equations over GF(2), as well as another equation which holds with high probability.

Equation finding

The question arises whether we have identified all quadratic equations in the w and x variables for the AES inversion operation. This question can be answered using linear algebra. For simplicity, we consider how to identify all bilinear forms in the w and x variables.

Suppose $a \in \mathbf{F}^*$, and let a_i and $\bar{a}_j$ denote the components of a and a^{-1} when considered as vectors of $GF(2)^8$. We can then define $\mathbf{a} \in GF(2)^{64}$ to be the vector

$$\mathbf{a} = (a_0\bar{a}_0, a_0\bar{a}_1, a_0\bar{a}_2, \ldots, a_7\bar{a}_6, a_7\bar{a}_7)^T .$$

If we have a bilinear equation

$$0 = \sum_{i=0}^{7} \sum_{j=0}^{7} c_{ij} w_i x_j$$

over GF(2) in the components of w and x of an AES inversion function, then let $\mathbf{c} = (c_{ij})$ denote the vector of bilinear equation coefficients. For any $a \in \mathbf{F}^*$, we have $\mathbf{a}^T \mathbf{c} = 0$ and we can construct a $255{\times}64$ matrix A, with rows given by the corresponding vectors $\mathbf{a}^T$, such that $A\mathbf{c} = 0$. The coefficients $\mathbf{c}$ for which the above bilinear equation holds are given by $\ker(A)$, and we can find all such bilinear equations. This technique can be extended to biaffine and general quadratic equations by extending the set of monomials.

Using this method we can show that we have identified all quadratic equations over GF(2) in the input and output of the AES inversion operation. More generally, this kernel technique can be extended to find polynomial equations between the input and output of cryptographic functions such as S-boxes by using an appropriate matrix A.

A sparse equation system

We are now in a position to give equation systems over GF(2) for the AES. The equation system described here is very sparse, in which the variables represent the input and output of the inversion operations and the round keys. We also require some auxiliary key variables in order to describe the key schedule.

	Source		
	State	*Key*	*Total*
Variables	2560	1728	4288

	Source		
	Encryption	*Key schedule*	*Total*
Field equations	2560	1728	4288
Linear equations	1408	1280	2688
Inversion equations	6400	1600	8000
Overall equations	10368	4608	14976

Figure 5.1. The number of variables and equations in a sparse quadratic equation system for the AES.

We first note that since the equation system is over $GF(2)$, any variable z satisfies the field equation $z^2 + z = 0$. In the equation system for an AES encryption, as well as the key variables there are 1280 inversion input variables w and 1280 inversion output variables x. These variables are used in $11 \cdot 128 = 1408$ linear equations and they give 2560 field equations. If we assume that encryption does not contain a 0-inversion, there are 160 inversions in an encryption with 40 quadratic equations for each inversion, giving 6400 quadratic equations in total.

In the AES key schedule, there are $11 \cdot 128 = 1408$ round key variables. The inversion function is only applied to four of the sixteen round key bytes at every round, so the key schedule uses 40 inversions. In order to describe the key schedule using sparse quadratic equations, we add the components of the output of the key schedule inversion as variables. Thus there are 320 inversion output variables, giving 1728 key variables in total. If there are no 0-inversions in the key schedule, we obtain $40 \cdot 40 = 1600$ quadratic equations, 1728 field equations, and $10 \cdot 128 = 1280$ linear equations in the key schedule.

This sparse quadratic equation system is summarised in Figure 5.1. We see that overall we obtain a system with 14976 equations in 4288 variables. We note that any equation system for the AES is somewhat arbitrary, and the choice of an equation system for the AES depends to some extent on the intended use of the system. For example, the quadratic equations for the AES inversion given in [31] are more complicated than the ones given in Appendix A, but the associated linear equations are much simpler. Furthermore, we can always use linear equations to eliminate variables by simple substitution so as to give a system with fewer variables, though generally at the cost of making the nonlinear equations more complex.

	Source		
	State	*Key*	*Total*
Variables	1280	320	1600

	Source		
	Encryption	*Key schedule*	*Total*
Field equations	1280	320	1600
Inversion equations	6400	1600	8000
Overall equations	7680	1920	9600

Figure 5.2. The number of variables and equations in a compact quadratic equation system for the AES.

A compact equation system

We can produce a more compact equation system by using linear equations in the AES system to substitute and eliminate variables. We first consider the linear relations in the encryption process. We let $\mathbf{w}_i$ and $\mathbf{x}_i$ denote the input to and the output from an AES inversion respectively, and we let $\mathbf{k}_i$ denote the round key and **63** the vector of repeated S-box constants. These are considered as vectors of length 128 over GF(2). For plaintext $\mathbf{p}$ and ciphertext $\mathbf{c}$, AES encryption is described by

$$
\begin{aligned}
\mathbf{w}_0 &= \mathbf{p} + \mathbf{k}_0 + \mathbf{63} \\
\mathbf{w}_i &= M\mathbf{x}_{i-1} + \mathbf{k}_i + \mathbf{63} && [i = 1, \ldots, 9] \\
\mathbf{c} &= M^*\mathbf{x}_9 + \mathbf{k}_{10} + \mathbf{63} \ ,
\end{aligned}
$$

where M^* is the modified matrix for the final round. We therefore write

$$
\begin{aligned}
\mathbf{x}_i &= M^{-1}(\mathbf{w}_{i+1} + \mathbf{k}_{i+1} + \mathbf{63}) && [i = 0, \ldots, 8] \\
\mathbf{x}_9 &= (M^*)^{-1}(\mathbf{k}_{10} + \mathbf{c} + \mathbf{63}) \ ,
\end{aligned}
$$

and eliminate the 1280 variables arising from the vectors $\mathbf{x}_i$ ($0 \leq i \leq 9$).

We now consider linear relations in the key schedule. Some related issues are considered in [2]. Let $\mathbf{s}_i$ denote the 32-bit output vector of the inversion operation in the AES key schedule. The output of the F-function of the key schedule is given by $Q\mathbf{s}_i + Q\mathbf{r}_i$, where Q is the 32×32 matrix over GF(2) corresponding to the byte rotation and $\mathbf{r}_i$ is a round constant vector of length 32. The relationship between successive round keys is given by

$$
\begin{pmatrix} \mathbf{k}_i^{(0)} \\ \mathbf{k}_i^{(1)} \\ \mathbf{k}_i^{(2)} \\ \mathbf{k}_i^{(3)} \end{pmatrix} = \begin{pmatrix} I & 0 & 0 & 0 \\ I & I & 0 & 0 \\ I & I & I & 0 \\ I & I & I & I \end{pmatrix} \begin{pmatrix} \mathbf{k}_{i-1}^{(0)} \\ \mathbf{k}_{i-1}^{(1)} \\ \mathbf{k}_{i-1}^{(2)} \\ \mathbf{k}_{i-1}^{(3)} \end{pmatrix} + \begin{pmatrix} Q(\mathbf{s}_{i-1} + \mathbf{r}_{i-1}) \\ Q(\mathbf{s}_{i-1} + \mathbf{r}_{i-1}) \\ Q(\mathbf{s}_{i-1} + \mathbf{r}_{i-1}) \\ Q(\mathbf{s}_{i-1} + \mathbf{r}_{i-1}) \end{pmatrix} ,
$$

where $\mathbf{k}_i^{(j)}$ denotes a vector of length 32 corresponding to a column of the round key array. Thus there exist a 128×128 matrix A and a 128×32 matrix B over GF(2) such that

$$\mathbf{k}_i = A\mathbf{k}_{i-1} + B\left(\mathbf{s}_{i-1} + \mathbf{r}_{i-1}\right) = A^i\mathbf{k}_0 + \sum_{j=1}^{i} A^{j-1}B\left(\mathbf{s}_{i-j} + \mathbf{r}_{i-j}\right).$$

If we use the relation $\mathbf{k}_0 = \mathbf{w}_0 + \mathbf{p} + \mathbf{63}$ from the encryption, we have

$$\mathbf{k}_i = A^i\mathbf{w}_0 + \sum_{j=1}^{i} A^{j-1}B\mathbf{s}_{i-j} + \sum_{j=1}^{i} A^{j-1}B\mathbf{r}_{i-j} + A^i(\mathbf{p} + \mathbf{63}),$$

and we can eliminate the $128 \cdot 11 = 1408$ variables that refer to the vectors $\mathbf{k}_i$ $(0 \le i \le 10)$.

We have used the linear equations in the AES system to eliminate $1280 + 1408 = 2688$ binary variables, and can now express the AES encryption in terms of $4288 - 2688 = 1600$ variables. These variables consist of the 1280 input variables w from the inversion operation in the encryption and the 320 key variables that are the output of the inversion operation in the key schedule.

This compact quadratic equation system is summarised in Figure 5.2, where we see that an AES encryption can be described by an equation system with 9600 quadratic equations in 1600 variables. Of these 9600 quadratic equations, 1600 are field equations and 8000 are from the inversion operations. We note that some of these quadratic equations are more complicated than others, and a sparser system can be obtained by considering the 4800 equations arising from the quadratic equations.

Equation system for the DES

For the purposes of comparison with the AES, we describe an equation system for the DES. We have already seen a quadratic equation for the first DES S-box at the beginning of this chapter. However such quadratic equations for a DES S-box are rare. There are only six other quadratic equations relating the input and output bits of a DES S-box: five for the fourth S-box and one for the fifth S-box [114]. Thus it is not possible to describe a DES encryption in terms of quadratic polynomials over GF(2) in the state variables.

We therefore consider cubic polynomials. There are ten binary variables for a DES S-box, so there are $\binom{10}{3} + \binom{10}{2} + \binom{10}{1} + \binom{10}{0} = 176$ monomials of degree at most three in these ten variables. We can find cubic polynomials satisfied by these ten variables using the equation finding kernel technique described earlier. This requires us to find the kernel of a 64×176 matrix over GF(2). Such a matrix for a DES S-box has a kernel of dimension at least $176 - 64 = 112$. We can thus fully describe a DES encryption using a cubic equation system in these variables, and the number of equations in such a system can be calculated using the figures given in [114]. We note that the nonlinear equations for the DES do not seem to possess any obvious structure, unlike those for the AES.

3. Equation Systems over $\mathbf{GF}(2^8)$

We now derive an equation system over the Rijndael field $\mathbf{F}$ to describe an AES encryption [89]. In Section 4.3 we introduced a block cipher called the Big Encryption System (BES) and showed that the AES could be represented as the BES with a restricted message space. The equation system over $\mathbf{F}$ that we give below is based on this representation. It is clearly much simpler than the equation system over GF(2) discussed in Section 4.2.

BES equation system

The state space of the BES is $\mathbf{B} = \mathbf{F}^{128}$ and the typical round function of the BES is given by

$$\mathbf{b} \mapsto M_B\left(\mathbf{b}^{(-1)}\right) + (\mathbf{k}_B)_i,$$

where M_B is the linear diffusion matrix given in Section 4.3 and $\mathbf{k}_{Bi}$ is a BES round key. Similarly to the AES, we denote the state vectors before and after the inversion operation by $\mathbf{w}_i \in \mathbf{B}$ and $\mathbf{x}_i \in \mathbf{B}$ $(0 \leq i \leq 9)$ respectively. The encryption of plaintext $\mathbf{p} \in \mathbf{B}$ to ciphertext $\mathbf{c} \in \mathbf{B}$ by the BES is then described by

$$
\begin{aligned}
\mathbf{w}_0 &= \mathbf{p} + \mathbf{k}_0 & \\
\mathbf{x}_i &= \mathbf{w}_i^{(-1)} & [i = 0,\ldots,9] \\
\mathbf{w}_i &= M_B\mathbf{x}_{i-1} + \mathbf{k}_i & [i = 1,\ldots,9] \\
\mathbf{c} &= M_B^*\mathbf{x}_9 + \mathbf{k}_{10},
\end{aligned}
$$

where M_B^* is the modified version of matrix M_B for the final round. If we denote the components of $\mathbf{x}_i$ by $x_{i,(j,m)}$ $(0 \leq j \leq 15$ and $0 \leq m \leq 7)$, we obtain the following system for a BES encryption

$$
\begin{aligned}
w_{0,(j,m)} &= p_{(j,m)} + k_{0,(j,m)} & \\
x_{i,(j,m)} &= w_{i,(j,m)}^{(-1)} & [i = 0,\ldots,9] \\
w_{i,(j,m)} &= (M_B\mathbf{x}_{i-1})_{(j,m)} + k_{i,(j,m)} & [i = 1,\ldots,9] \\
c_{(j,m)} &= (M_B^*\mathbf{x}_9)_{(j,m)} + k_{10,(j,m)}.
\end{aligned}
$$

Under the assumption that 0-inversions do not occur as part of the encryption (Section 3.2), this equation system becomes

$$
\begin{aligned}
0 &= w_{0,(j,m)} + p_{(j,m)} + k_{0,(j,m)} \\
0 &= x_{i,(j,m)} w_{i,(j,m)} + 1 & [i = 0, \ldots, 9] \\
0 &= w_{i,(j,m)} + (M_B x_{i-1})_{(j,m)} + k_{i,(j,m)} & [i = 1, \ldots, 9] \\
0 &= c_{(j,m)} + (M_B^* x_9)_{(j,m)} + k_{10,(j,m)}.
\end{aligned}
$$

AES encryption embedded in the BES

An AES encryption can be embedded in the BES, so the above equation system for a BES encryption also describes an embedded AES encryption. However, the embedded state variables of an AES encryption are elements of $\mathbf{B_A}$, the AES subset of $\mathbf{B}$, and so possess the conjugacy property. This conjugacy property gives us further multivariate quadratic equations. Thus embedding an AES encryption in the BES gives the equation system over $\mathbf{F}$ (where $m + 1$ is interpreted modulo 8)

$$
\begin{aligned}
0 &= w_{0,(j,m)} + p_{(j,m)} + k_{0,(j,m)} \\
0 &= w_{i,(j,m)} + k_{i,(j,m)} + (M_B x_{i-1})_{(j,m)} & [i = 1, \ldots, 9] \\
0 &= c_{(j,m)} + k_{10,(j,m)} + (M_B^* x_9)_{(j,m)} \\
0 &= x_{i,(j,m)} w_{i,(j,m)} + 1 & [i = 0, \ldots, 9] \\
0 &= x_{i,(j,m)}^2 + x_{i,(j,m+1)} & [i = 0, \ldots, 9] \\
0 &= w_{i,(j,m)}^2 + w_{i,(j,m+1)} & [i = 0, \ldots, 9].
\end{aligned}
$$

Similarly, we can obtain an equation system for the AES key schedule embedded in the BES. This key schedule equation system has 1408 round key variables that are also used in the encryption equations and 320 auxiliary variables, which are the output variables for inversion in the key schedule.

The BES gives an equation system for the AES over $\mathbf{F}$. Every equation and variable has a counterpart in the equation system over GF(2) for the AES (Section 5.2). Figures 5.1 and 5.2 thus describe the number of equations and variables for AES equation systems over $\mathbf{F}$. We note however that the equation systems for the AES over $\mathbf{F}$ are extremely sparse compared with the corresponding equation systems over GF(2). In particular, every quadratic equation in the sparse system of Figure 5.1 has only one nonlinear term. Furthermore, every quadratic equation in the compact system of Figure 5.2 is much simpler than the corresponding equation over GF(2).

Small scale example

We illustrate an equation system over $\mathbf{F}$ by considering small scale variants of the AES (Section 3.3). We give the entire equation system over

$GF(2^4)$ in Appendix C for the small scale variant $SR(2, 2, 2, 4)$, which has two rounds and a 2×2 state array of elements of $GF(2^4)$. This equation system is analogous to the equation system over $\mathbf{F}$ for the AES.

4. Gröbner Basis Equation System

We now show how to obtain a different system of polynomial equations over the Rijndael field $\mathbf{F}$. We follow the approach of [14] and consider the non-linear operation in the S-box as $x \mapsto x^{254}$ rather than inversion in the Rijndael field $\mathbf{F}$. Such systems contain equations that are denser and have higher degree than those described in Section 5.3, and have also been considered in [116]. However, the equation system of [14] does show some interesting algebraic properties of the AES encryption.

We let $\mathbf{w}_i = (w_{i,0}, \ldots, w_{i,15}) \in \mathbf{F}^{16}$ denote the round input ($0 \leq i \leq 9$) and $\mathbf{k}_i = (k_{i,0}, \ldots, k_{i,15})$ the round key ($0 \leq i \leq 10$). Furthermore, we let $\mathbf{S}(\mathbf{w}_i) = (g(w_{i,0}), \ldots, g(w_{i,15}))$ denote the output of the SubBytes operation, where the polynomial $g(z)$ is the interpolating polynomial for the S-box and is given by

$$05z^{254} + 09z^{253} + \text{F9}z^{251} + 25z^{247} + \text{F4}z^{239} + 01z^{223} + \text{B5}z^{191} + 8\text{F}z^{127} + 63.$$

If $\mathbf{p}$ and $\mathbf{c}$ denote the plaintext and ciphertext respectively, then an AES encryption is given by

$$
\begin{aligned}
\mathbf{w}_0 &= \mathbf{p} + \mathbf{k}_0 \\
\mathbf{w}_i &= \overline{C}\,\overline{R}\,(\mathbf{S}(\mathbf{w}_{i-1})) + \mathbf{k}_i \qquad [i = 1, \ldots, 9] \\
\mathbf{c} &= \overline{R}\,(\mathbf{S}(\mathbf{w}_9)) + \mathbf{k}_{10},
\end{aligned}
$$

where $\overline{R}$ and $\overline{C}$ are the 16×16 matrices over $\mathbf{F}$ corresponding to the ShiftRows and MixColumns operations (Section 4.1).

We can now rearrange the system to obtain

$$
\begin{aligned}
0 &= \mathbf{w}_0 + \mathbf{k}_0 + \mathbf{p} \\
0 &= \mathbf{S}(\mathbf{w}_{i-1}) + \left(\overline{C}\,\overline{R}\right)^{-1}(\mathbf{w}_i + \mathbf{k}_i) \qquad [i = 1, \ldots, 9] \\
0 &= \mathbf{S}(\mathbf{w}_9) + \overline{R}^{-1}(\mathbf{k}_{10} + \mathbf{c}).
\end{aligned}
$$

This gives an equation system with 176 equations, of which 16 equations are linear and the other 160 equations each have total degree 254.

We can perform a similar rearrangement with the key schedule equations using the inverse S-box, though its interpolating polynomial is dense. The coefficients of this polynomial $h(z)$ are given in Figure 5.3, where

$$h(z) = 05z^{254} + \text{CF}z^{253} + \ldots + \text{F3}z + 52.$$

05	CF	B3	16	55	C0	7A	01	22	D8	6B	A6	1F	0D	BC
49	85	B4	1B	5E	BD	18	1D	6D	C5	23	09	43	68	80
6C	CC	42	9F	0F	D2	3B	2C	5F	BE	AE	E4	93	8B	CB
65	C0	1E	8E	32	1D	A5	76	A9	2C	13	05	60	FD	1B
AB	64	C1	A8	7F	55	DB	EC	20	C4	DB	7E	92	80	A3
59	91	91	81	4E	11	DD	4E	D3	E3	19	E7	03	24	45
DA	EA	87	2D	23	82	38	B7	9E	B3	2A	3E	1C	EC	C3
45	ED	D5	2A	8D	ED	37	26	E0	BC	58	E2	6C	24	55
C7	AA	09	4F	82	CA	10	EE	1A	2E	40	27	81	92	B1
02	8B	87	7F	B0	6F	53	08	CB	03	B0	DF	1F	A7	A2
FE	8E	A8	E1	71	FF	55	5A	1D	9D	BF	E8	BA	6B	72
E3	04	D9	38	D3	B9	16	52	18	19	3E	9E	03	56	A6
71	03	E4	86	F5	B0	05	D1	10	E2	E5	CB	B1	F2	8E
C7	0C	A7	BF	46	0B	01	C5	A3	50	77	EA	05	65	8E
89	D4	6D	D3	75	65	13	2F	86	AF	7C	7B	85	C8	E8
04	7B	CF	2F	8A	9A	3D	CF	21	39	D9	29	73	F6	23
40	1B	B2	C0	6D	85	1C	8A	2C	BB	90	1E	7E	F3	52

Figure 5.3. Coefficients of the interpolating polynomial for the inverse S-box.

Thus we obtain $(1 \leq i \leq 10)$

$$
\begin{pmatrix} 0 \\ 0 \\ 0 \\ 0 \\ 0 \\ \vdots \\ 0 \end{pmatrix} = \begin{pmatrix} h(k_{i,0} + k_{(i-1),0} + \theta^{i-1}) \\ h(k_{i,1} + k_{(i-1),1}) \\ h(k_{i,2} + k_{(i-1),2}) \\ h(k_{i,3} + k_{(i-1),3}) \\ k_{i,4} + k_{(i-1),4} \\ \vdots \\ k_{i,15} + k_{(i-1),15} \end{pmatrix} + \begin{pmatrix} k_{i-1,15} \\ k_{i-1,12} \\ k_{i-1,13} \\ k_{i-1,14} \\ k_{i,0} \\ \vdots \\ k_{i,11} \end{pmatrix}.
$$

We have thus constructed an equation system over $\mathbf{F}$ for an AES encryption in 336 variables. This equation system comprises 176 polynomial equations arising from the encryption operation and 160 from the key schedule. Of these 336 equations, 200 equations each have total degree 254, while the remaining 136 equations are linear.

We can also consider this system as a set of polynomials in the multivariate polynomial ring

$$
\mathbf{F}[w_{0,0}, \ldots, w_{0,15}, k_{0,0}, \ldots, k_{10,15}, w_{1,0}, \ldots, w_{9,15}]
$$

with 336 variables over $\mathbf{F}$. We consider this ring under the *glex* monomial ordering (Section 2.2), where the variables are ordered as

$$
w_{0,0} \prec \ldots \prec w_{0,15} \prec k_{0,0} \prec \ldots \prec k_{10,15} \prec w_{1,0} \prec \ldots \prec w_{9,15}.
$$

Under this ordering, the 160 polynomials of degree 254 derived from the encryption operation have $w_{i,j}^{254}$ as their leading monomial. The remaining 16 linear equations are those containing the plaintext and have $k_{0,j}$ as their leading monomial. In the key schedule, the linear equations have $k_{i,j}$ as their leading monomial ($1 \leq i \leq 10$ and $4 \leq j \leq 15$), whilst the nonlinear equations have $k_{i,j}^{254}$ as their leading monomial ($1 \leq i \leq 10$ and $0 \leq j \leq 3$). Thus the leading monomials of all polynomials are pairwise coprime, and Theorem 5.1 now follows immediately from Theorem 2.80.

THEOREM 5.1 The set of polynomials over $\mathrm{GF}(2^8)$ derived from the AES encryption as above is a Gröbner basis with respect to the *glex* monomial ordering.

Some consequences of Theorem 5.1 are explored in Section 6.3.

Chapter 6

ANALYSIS OF AES EQUATION SYSTEMS

After Rijndael was adopted as the AES, the possibility of algebraic attacks led to much speculation [75, 110, 112]. This might be seen as part of a growing interest in the wider application of computational algebra to cryptography. Systems of multivariate polynomial equations have been proposed in asymmetric cryptology [100] and the analysis of some cryptosystems, most notably certain stream ciphers [29], demonstrate the importance of computational algebra techniques.

Solving systems of multivariate polynomial equations is a classical problem in algebraic geometry and computer algebra [33, 34]. Suppose we have a field $\mathbb{F}$ and a multivariate polynomial ring $\mathbb{F}[x_1, \ldots, x_n]$ in n variables over $\mathbb{F}$. Given a set of m polynomials f_1, ..., f_m in $\mathbb{F}[x_1, \ldots, x_n]$, we might wish to find solutions to the equation system $f_i = 0$ $(1 \leq i \leq m)$, that is to find $(a_1, \ldots, a_n) \in \mathbb{F}^n$ such that

$$f_1(a_1, \ldots, a_n) = \ldots = f_m(a_1, \ldots, a_n) = 0.$$

This problem is equivalent to finding the affine variety associated with the ideal $I = \langle f_1, \ldots, f_m \rangle \lhd \mathbb{F}[x_1, \ldots, x_n]$ generated by the polynomials $f_1, \ldots, f_m$ (Section 2.5). This affine variety $\mathcal{V}(I)$ is defined by

$$\mathcal{V}(I) = \left\{ (a_1, a_2, \ldots, a_n) \in \mathbb{F}^n \mid f(a_1, a_2, \ldots, a_n) = 0 \quad \text{for all } f \in I \right\}.$$

A common technique to obtain the set of solutions of a polynomial system is to compute the reduced Gröbner basis of the ideal I, particularly with respect to the *lex* monomial ordering (Section 2.2). By finding such a Gröbner basis of I, we can obtain all solutions to an equation system in the algebraic closure of $\mathbb{F}$. However, there are situations where the solutions of a polynomial system can be found without calculating the reduced Gröbner basis of I.

The problem of solving systems of multivariate equations over a finite field is known to be NP-hard [54]. However we do not expect the problem to be so hard on average. Furthermore, in the cases of interest in cryptology, the systems often have some special properties. For example, the field $\mathbb{F}$ is often a finite field of characteristic 2 and the solutions sought usually lie in $\mathbb{F}$. A common technique in this case is to add the finite field relations $x_i^q - x_i$ ($1 \leq i \leq n$) to the existing set of equations, where q is the order of $\mathbb{F}$. This gives a set of $m+n$ equations and ensures that all found solutions lie in $\mathbb{F}$.

Equation systems that occur in symmetric cryptology often contain many more equations than variables. We call this type of equation system an *overdefined* or *overdetermined* system. Overdefined systems are often easier to solve. In fact, the very existence of an overdefined multivariate quadratic equation system for the AES was the basis for much of the early research into algebraic attacks against the cipher [31, 32].

Various methods have been suggested for the analysis of such equation systems, and there is much literature on the subject [33, 34, 66, 72]. We discuss some proposed methods of solution for such equation systems. We give an overview of some of the classical methods, such as Buchberger's algorithm for computing a Gröbner basis, as well as methods that have been specifically proposed in the context of cryptology, such as the XL method. We also discuss the applicability of other methods designed to exploit the structure of the AES equations.

1.   Gröbner Basis Methods

Gröbner basis algorithms are well-known general purpose methods for solving systems of multivariate polynomial equations. Most computer algebra packages, such as the MAGMA, SINGULAR and MAPLE packages, include implementations of Gröbner basis algorithms. This is often the default technique for computing the solution of a system of polynomial equations. The classical general algorithm for computing a Gröbner basis of a polynomial ideal is Buchberger's algorithm [13].

Buchberger's algorithm

We consider the polynomial ring $\mathbb{F}[x_1, \ldots, x_n]$ with a monomial ordering. Suppose $I \lhd \mathbb{F}[x_1, \ldots, x_n]$ is an ideal of this polynomial ring with a basis $F = \{f_1, \ldots, f_m\}$. We can define the *S-polynomial* of any pair of polynomials (f_i, f_j) of F by

$$S(f_i, f_j) = \left(\frac{\operatorname{lcm}(\operatorname{LM}(f_i), \operatorname{LM}(f_j))}{\operatorname{LT}(f_i)} \right) f_i - \left(\frac{\operatorname{lcm}(\operatorname{LM}(f_i), \operatorname{LM}(f_j))}{\operatorname{LT}(f_j)} \right) f_j,$$

```
1:  Input: F = {f_1, ..., f_m}
2:  Output: Gröbner Basis G = {g_1, ..., g_s} for the ideal generated by F
3:
4:  G := F
5:  repeat
6:      G' := G
7:      for each pair of distinct polynomials p and q in G' do
8:          Construct the S-polynomial S(p, q)
9:          Compute the remainder r of division of S(p, q) by the polynomials in G'
10:         if r ≠ 0 then
11:             G := G ∪ {r}
12:         end if
13:     end for
14: until G = G'
15:
16: return G
```

Figure 6.1. Buchberger's algorithm for computing a Gröbner basis.

with lcm being the *least common multiple*. The S-polynomial $S(f_i, f_j)$ is a polynomial in the ideal I which is computed essentially by cancelling the leading terms of the two polynomials f_i and f_j. Buchberger's algorithm uses S-polynomials and Theorem 6.1 to compute a Gröbner basis of the ideal I.

THEOREM 6.1 Let $\mathbb{F}[x_1, \ldots, x_n]$ be a polynomial ring with a monomial ordering and let I be an ideal of $\mathbb{F}[x_1, \ldots, x_n]$. A basis $G = \{f_1, ..., f_m\}$ for the ideal I is a Gröbner basis for I if and only if every S-polynomial $S(f_i, f_j)$ of pairs of distinct polynomials $f_i, f_j \in G$ has remainder 0 upon division (reduction) by G.

Buchberger's algorithm [33] is given in Figure 6.1. The algorithm can be modified to perform the autoreduction of the set G as the last step of the algorithm. This modified algorithm computes the unique reduced Gröbner basis of the ideal $I = \langle f_1, ..., f_m \rangle$. Such a reduced Gröbner basis, especially with respect to the *lex* ordering, can be used to compute solutions of the equation system

$$f_1(x_1, \ldots, x_n) = 0, \ \ldots \ , f_m(x_1, \ldots, x_n) = 0.$$

EXAMPLE 6.2 We consider the polynomial ring $\mathbb{C}[x, y]$ of polynomials in two variables over the complex numbers with the *lex* ordering (with $y \prec x$). Let $I \lhd \mathbb{C}[x, y]$ be the ideal generated by the two polynomials

$$f_1 = x^2 y - 1 \text{ and } f_2 = xy^2 - x.$$

We compute the Gröbner basis of this ideal I by using Buchberger's algorithm. We initially set $G = \{f_1, f_2\}$ and compute the S-polynomial of the two polynomials in G, so

$$\begin{aligned} S(f_1, f_2) &= \tfrac{x^2y^2}{x^2y}(x^2y - 1) - \tfrac{x^2y^2}{xy^2}(xy^2 - x) \\ &= y(x^2y - 1) - x(xy^2 - x) = x^2 - y. \end{aligned}$$

We note that the leading term of $S(f_1, f_2)$ is x^2, whilst $\mathrm{LT}(f_1) = x^2y$ and $\mathrm{LT}(f_2) = xy^2$. Thus $S(f_1, f_2)$ is a polynomial that cannot be reduced by either polynomial in $G = \{f_1, f_2\}$. We thus set $f_3 = S(f_1, f_2) = x^2 - y$ and include f_3 in G to obtain $G = \{f_1, f_2, f_3\}$.

We now compute the S-polynomial of f_1 and f_3 to obtain

$$\begin{aligned} S(f_1, f_3) &= \tfrac{x^2y}{x^2y}(x^2y - 1) - \tfrac{x^2y}{x^2}(x^2 - y) \\ &= (x^2y - 1) - y(x^2 - y) = y^2 - 1. \end{aligned}$$

We note that $S(f_1, f_3)$ cannot be reduced by the set $G = \{f_1, f_2, f_3\}$. We thus set $f_4 = S(f_1, f_3) = y^2 - 1$ and include f_4 in G to obtain $G = \{f_1, f_2, f_3, f_4\}$.

We next compute $S(f_2, f_3) = -x^2 + y^3$. We note that its leading term is $-x^2$, which is divisible by $\mathrm{LT}(f_3) = x^2$. We can thus divide $S(f_2, f_3)$ by f_3 and then by f_4 to obtain

$$S(f_2, f_3) = -x^2 + y^3 = -f_3 + (y^3 - y) = -f_3 + yf_4.$$

Similarly, we have $S(f_1, f_4) = f_3$, $S(f_2, f_4) = 0$ and $S(f_3, f_4) = f_3 - yf_4$. Thus the S-polynomials of all pairs of polynomials in G are reduced to 0 by G, and so $G = \{f_1, f_2, f_3, f_4\}$ is a Gröbner basis of I.

We can now reduce the set G to obtain the reduced Gröbner basis $\{x^2 - y, y^2 - 1\}$ of I. It follows that the equation system

$$x^2 - y = 0 \text{ and } y^2 - 1 = 0$$

has the same solution set as the equation system

$$x^2y - 1 = 0 \text{ and } xy^2 - x = 0.$$

Thus we have the complete solution set $\{(1, 1), (-1, 1), (i, -1), (-i, -1)\}$ in the complex numbers for the system $x^2y - 1 = xy^2 - x = 0$. □

Buchberger's algorithm can be thought of as a generalisation of the Euclidean algorithm for calculating the greatest common divisor of a set of univariate polynomials. The univariate polynomial ring $\mathbb{F}[x]$ is a principal ideal domain. Thus the ideal $\langle f_1, \ldots, f_m \rangle$ is generated by the polynomial $g = \gcd(f_1, \ldots, f_m)$ and g is just the reduced Gröbner basis of $\langle f_1, \ldots, f_m \rangle$. This basis is calculated by Buchberger's algorithm.

Buchberger's algorithm can also be thought of as a generalisation of Gaussian reduction to nonlinear polynomials. The reduced Gröbner basis of an ideal generated by a set of linear polynomials is a set of linear polynomials in echelon form. The equivalent echelon set of linear polynomials is thus calculated by Buchberger's algorithm.

As presented in Figure 6.1, Buchberger's algorithm can be used directly to compute the Gröbner basis of a polynomial ideal. However, it is not particularly efficient as many of the S-polynomials generated in the first step of the algorithm reduce to zero. Thus many unnecessary reductions are performed, and reductions are by far the most computationally intensive part of the algorithm. A good way to improve the efficiency of the algorithm is to identify those pairs of polynomials whose S-polynomials are known (before calculation) to reduce to zero. Such pairs of polynomials can be identified using *Buchberger's criteria*. For example, *Buchberger's first criterion* is a consequence of Theorem 6.3.

THEOREM 6.3 If $f_1, f_2 \in \mathbb{F}[x_1, \ldots, x_n]$ are two polynomials with coprime leading monomials, then the S-polynomial $S(f_1, f_2)$ reduces to zero with respect to $\{f_1, f_2\}$. Thus Buchberger's algorithm does not need to consider $S(f_1, f_2)$.

We note that Theorem 2.80 follows immediately from Theorem 6.3. Such criteria can be used to modify and improve the efficiency of Buchberger's algorithm [33].

Complexity of Buchberger's algorithm

Even when Buchberger's criteria are incorporated, the algorithm still performs many unnecessary S-polynomial reductions. Further issues, such as the order in which S-polynomials are processed or the choice of monomial ordering, also have a strong influence on the efficiency of the algorithm. For example, the *grevlex* ordering is often the most efficient monomial ordering.

The complexity of Buchberger's algorithm is closely related to the total degree of the intermediate polynomials generated by the algorithm. There are examples where the computation of a Gröbner basis of an ideal generated by polynomials of degree at most d involves polynomials of degree proportional to 2^{2^d} [33]. In fact, Buchberger's algorithm can have double exponential complexity. There are also examples for which the computation of a Gröbner basis using Buchberger's algorithm requires an enormous amount of time and memory. However, these examples tend to be somewhat artificial and, in general, the running time and storage requirements of Buchberger's algorithm seem to be much more manageable for generic cases [5, 79].

The required solutions for AES equation systems lie in the ground field. Thus we usually include field equations of the form $z_i^2 + z_i = 0$ when considering the AES equation system over $GF(2)$ (Section 5.2). For the equation system for the AES over $GF(2^8)$ given by the BES, we would include the conjugacy equations of the form $z_i^2 + z_{i+1} = 0$ (Section 5.3). For such systems, the degree of the intermediate polynomials generated by Buchberger's algorithm is at most the total number of variables in the equation system. Thus the complexity of Buchberger's algorithm in relation to an AES equation system is at worst single exponential. However, it is very unlikely that Buchberger's algorithm without further optimizations could be used to find the solution of the type of equation systems arising from established block ciphers.

F4 and F5 algorithms

The F4 and F5 algorithms have been proposed as alternative approaches for computing Gröbner bases [46, 47]. The F4 algorithm can be considered an enhanced version of Buchberger's algorithm. Since the main computational cost of Buchberger's algorithm lies in polynomial reductions, which take place sequentially, the F4 algorithm essentially replaces many sequential polynomial reductions with a matrix reduction. This can potentially give a faster algorithm than Buchberger's algorithm. The F5 algorithm works in similar manner, using ideas introduced in [72]. The F5 algorithm also includes an optimal criterion that ensures that, under some conditions, all the matrices generated are full-rank.

The idea of combining Gröbner basis computation with Gaussian elimination was first discussed in [72]. The F4 and F5 algorithms are based on this idea and work by performing the multivariate division algorithm as a matrix reduction. We illustrate this idea in Example 6.4 [115].

EXAMPLE 6.4 We consider the polynomial ring $\mathbb{R}[x, y, z]$ of polynomials in three variables over the real numbers with the *lex* ordering. Suppose we wish to reduce the following polynomials

$$f_1 = 3x^3yz - 5xy \text{ and } f_2 = 5x^2z^2 + 3xy + 1$$

by the (ordered) set of polynomials $\{g_1, g_2\}$, where

$$g_1 = xy - 2z \text{ and } g_2 = x^2z - 3yz.$$

This is a typical operation that is required by Buchberger's algorithm. We would first reduce f_1 with respect to $\{g_1, g_2\}$. The reduction steps required can be seen from the series of equalities

$$
\begin{aligned}
f_1 &= 3x^3yz - 5xy \\
&= 6x^2z^2 - 5xy && +(3x^2z)g_1 \\
&= -5xy + 18yz^2 && +(3x^2z)g_1 + (6z)g_2 \\
&= 18yz^2 - 10z && +(3x^2z)g_1 + (6z)g_2 - (5)g_1.
\end{aligned}
$$

Thus f_1 reduces to $18yz^2 - 10z$ with respect to $\{g_1, g_2\}$. We would then reduce f_2 with respect to $\{g_1, g_2\}$ by the steps indicated by

$$
\begin{aligned}
f_2 &= 5x^2z^2 + 3xy + 1 \\
&= 3xy + 15yz^2 + 1 && +(5z)g_2 \\
&= 15yz^2 + 6z + 1 && +(5z)g_2 + (3)g_1.
\end{aligned}
$$

Thus f_2 reduces to $15yz^2 + 6z + 1$ with respect to $\{g_1, g_2\}$.

Buchberger's algorithm would perform these two reductions sequentially. However, the individual reduction steps for both f_1 and f_2 only require reduction with respect to $(x^2z)g_1$, g_1 and $(z)g_2$. The idea behind the F4 and F5 algorithms is to carry out these reduction operations as a matrix reduction. Thus we would construct the matrix of coefficients

$$
\begin{array}{c c}
& \begin{array}{cccccc} x^3yz & x^2z^2 & yz^2 & xy & z & 1 \end{array} \\
\begin{array}{c} f_1 \\ f_2 \\ x^2z\,g_1 \\ 1\,g_1 \\ z\,g_2 \end{array} &
\left(\begin{array}{cccccc}
3 & 0 & 0 & -5 & 0 & 0 \\
0 & 5 & 0 & -3 & 0 & 1 \\
1 & -2 & 0 & 0 & 0 & 0 \\
0 & 0 & 0 & 1 & -2 & 0 \\
0 & 1 & -3 & 0 & 0 & 0
\end{array}\right).
\end{array}
$$

The polynomials f_1 and f_2 give the first two rows. The polynomials $(x^2z)g_1$, g_1 or $(z)g_2$, which are required to perform the reduction, give the remaining three rows. The reduction steps correspond to the row reduction of the upper two rows using the lower three rows. Such a row reduction would give the following matrix

$$
\begin{array}{c c}
& \begin{array}{cccccc} x^3yz & x^2z^2 & yz^2 & xy & z & 1 \end{array} \\
\begin{array}{c} f_1 \\ f_2 \\ x^2z\,g_1 \\ 1\,g_1 \\ z\,g_2 \end{array} &
\left(\begin{array}{cccccc}
0 & 0 & 18 & 0 & -10 & 0 \\
0 & 0 & 15 & 0 & 6 & 1 \\
1 & -2 & 0 & 0 & 0 & 0 \\
0 & 0 & 0 & 1 & -2 & 0 \\
0 & 1 & -3 & 0 & 0 & 0
\end{array}\right).
\end{array}
$$

The first two rows of this row-reduced matrix give the reduction of f_1 and f_2 with respect to $\{g_1, g_2\}$, thereby giving the same result as the sequential reduction of Buchberger given above. $\qquad\square$

Complexity of F4 and F5

The F4 and F5 algorithms incrementally construct matrices similar to that given in Example 6.4. These matrices are used to compute the reduction of many polynomials simultaneously by computing the equivalent row-reduced matrix. This means that the F4 and F5 algorithms are potentially faster than Buchberger's algorithm. The F5 algorithm has the further advantage that usually only full-rank matrices are generated. This avoids further unnecessary reductions. Furthermore, the matrix operations used by the F4 and F5 algorithms can often be speeded up using specialised techniques, such as sparse matrix methods. Similarly to Buchberger's algorithm, the F4 and F5 algorithms require an efficient selection criterion in order to decide how many and which polynomials are used to construct the matrices. The efficiency of the F4 and F5 algorithms is highly dependent on this selection criterion.

The F4 and F5 algorithms are currently the fastest known general algorithms for computing Gröbner bases. However, it is not easy to implement these algorithms in a way that is efficient for all inputs. Furthermore, the F4 and F5 algorithms often require more memory than Buchberger's algorithm. The F4 algorithm is currently the default Gröbner basis algorithm in the computer algebra package MAGMA [77]. Both the F4 and F5 algorithms have been successfully used to solve a well-known cryptographic challenge [49, 115].

As with Buchberger's algorithm, it is not an easy task to estimate the complexity of the F4 algorithm for generic cases. However, the F5 algorithm can be shown to have complexity of the order of N_D^ω field operations, where N_D is the size of the largest matrix containing polynomials up to degree D that is constructed by the algorithm and ω ($2 < \omega \le 3$) is the exponent of matrix reduction (Definition 2.53). In general, the degree D of the generated polynomials is a critical parameter in the efficiency of both algorithms.

The complexity of Gröbner basis computations using the F5 algorithm is considered in [6], where upper bounds for the size of matrices generated and for the algorithm complexity for generic systems of quadratic equations over $GF(2)$ with m equations and n variables are given. For large quadratic systems with the same number of equations and variables ($n = m$), the maximum degree D is expected to be about $0.09n$ (asymptotically). This implies that the sizes of the matrices generated are exponential in the number of variables, and so the complexity of the F5 algorithm should also be exponential. Figure 6.2 gives some more general results [6].

The estimates of [6] are for generic systems, that is systems with no particular structure. By contrast, the equation systems arising in cryp-

Condition	Complexity of the F5 algorithm
m grows linearly with n	Exponential in n
$n \ll m \ll n^2$	Subexponential in n
m grows linearly with n^2	Polynomial in n

Figure 6.2. The asymptotic complexity of the F5 algorithm for generic quadratic systems with m equations in n variables over GF(2).

tology are typically very structured. Such structure often means that the matrices generated by the F4 and F5 algorithms are much smaller than for generic cases. This was the case for the system of quadratic equations over GF(2) with 80 equations and variables arising from the (80-bit) HFE asymmetric cryptosystem [100]. The estimates of [6] for the generic case predict that the maximum degree D of polynomials required to solve the HFE equation system would be 12. However, the maximum degree obtained in reality is 4. This implies that an HFE equation system is far easier to solve than a comparable generic system, allowing the HFE Challenge I to be broken [49].

The complexity of solving block cipher equation systems using the F5 algorithm is also considered in [6]. For a generic equation system comparable to the AES system, the maximum degree of the polynomials generated by the F4 and F5 algorithms is expected to be 69. Thus the matrix that would be generated by the F5 algorithm would have size about 2^{341}. This means that the complexity of the F5 algorithm in solving this generic system would be of the order of $2^{341\omega}$ field operations, where ω is the exponent of matrix reduction (Definition 2.53). The solution of a generic system of this size would clearly be intractable using these Gröbner basis algorithms. However, these estimates are for generic equation systems whereas the AES equation systems are highly structured. State and key variables generally only occur in equations with the state and key variables from neighbouring rounds. Such equation systems are clearly different from comparable generic equation systems, and Gröbner basis algorithms may exploit such structure and reduce the complexity of the computations.

Gröbner basis conversion

A Gröbner basis is computed with respect to a specific monomial ordering. It is often the case that a polynomial ideal has different (reduced) Gröbner bases for different monomial orderings. However, there are certain orderings that are particularly useful for obtaining the solution of a system of polynomial equations associated with an ideal. Such monomial orderings are called *elimination orderings*, and the *lex* ordering is the most well-known example of an elimination ordering. The usefulness of the *lex* ordering for solving systems of polynomial equations is given by the following classical result from Elimination Theory [33].

THEOREM 6.5 Let $I \lhd \mathbb{F}[x_1, \ldots, x_n]$ be a polynomial ideal and G be a Gröbner basis of I with respect to the *lex* ordering with $x_n \prec \ldots \prec x_1$. Then for every $0 \leq k < n$, the set

$$G_k = G \cap \mathbb{F}[x_{k+1}, \ldots, x_n]$$

is a Gröbner basis of the ideal $I_k = I \cap \mathbb{F}[x_{k+1}, \ldots, x_n]$.

Theorem 6.5 states that we can use a Gröbner basis with respect to the *lex* ordering to essentially eliminate variables from the polynomial equation system. In particular, if the set of univariate polynomials in x_n in the ideal I is non-empty, then this set is a principal ideal. The single element of the Gröbner basis $G_{n-1} = G \cap \mathbb{F}[x_n]$ is the generator of this principal ideal. The associated univariate equation corresponding to this generator can be solved for x_n. This process can be repeated sequentially so that we obtain solutions of the equation system one variable at a time.

A Gröbner basis of an ideal I with respect to an elimination ordering is therefore a particularly useful tool for solving a multivariate equation system. However, there are other monomial orderings that might lead to more efficient computations. While a Gröbner basis of the ideal I with respect to such a monomial ordering might not immediately give solutions to the equation system, one possible approach is to obtain a Gröbner basis with respect to an efficient ordering and to convert this into another Gröbner basis with respect to an elimination ordering.

There are algorithms that convert a Gröbner basis of $I \lhd \mathbb{F}[x_1, \ldots, x_n]$ with respect to one monomial ordering into a Gröbner basis of I with respect to another monomial ordering. If the ideal I is such that the quotient ring $R = \mathbb{F}[x_1, \ldots, x_n]/I$ has finite dimension as a vector space over $\mathbb{F}$, we say that I is a *zero-dimensional ideal*. In this case, the *FGLM algorithm* can be used to perform the conversion between two Gröbner bases of I with respect to different monomial orderings [48]. The FGLM algorithm uses techniques from linear algebra in the vector space R, and its complexity is given by Theorem 6.6 [14, 48].

THEOREM 6.6 Let $I \lhd \mathbb{F}[x_1, \ldots, x_n]$ be a zero-dimensional polynomial ideal such that the quotient ring $\mathbb{F}[x_1, \ldots, x_n]/I$ has finite dimension d as a vector space over $\mathbb{F}$. The FGLM algorithm can convert a Gröbner basis of I with respect to one monomial ordering into a Gröbner basis of I with respect to another monomial ordering with complexity of the order of nd^3 field operations.

The FGLM algorithm is often used in conjunction with Gröbner Basis algorithms to compute the solution of a system of polynomial equations associated to a zero-dimensional polynomial ideal. We first compute a Gröbner basis of the associated ideal I with respect to an efficient ordering, such as *grevlex*. We then use the FGLM algorithm to convert this Gröbner basis to a Gröbner basis for I with respect to an elimination ordering, such as *lex*. This new Gröbner basis for the associated ideal I then allows us to compute the solutions to the equation system.

We note that when I is not a zero-dimensional ideal, there are other methods for conversion of Gröbner bases, such as the *Gröbner Walk* [34]. However, the complexity of the Gröbner Walk depends on the particular orderings used, and little is known in general about the time and space requirements for the algorithm.

A Gröbner basis for the AES

In Section 5.4, we show how to represent an AES encryption by a particular set of polynomials over the Rijndael field $\mathbf{F}$. These formed a Gröbner basis G with respect to a very specific monomial ordering (Theorem 5.1). However, this particular monomial ordering is not useful for directly obtaining the solutions of the AES equation system. A possible approach is to convert this Gröbner basis G to another Gröbner basis G' with respect to the *lex* ordering.

Let $\mathbf{F}[x_{ij}, k_{rs}]$ be the polynomial ring in the encryption and key variables and $I = \langle G \rangle$ be the ideal generated by the Gröbner basis G defined in Section 5.4. It follows from Theorem 5.1 that the quotient ring $R = \mathbf{F}[x_{ij}, k_{rs}]/I$ has dimension $254^{200} \approx 2^{1598}$, so I is a zero-dimensional ideal, and the FGLM algorithm could, in principle, be used to convert between Gröbner bases for I. However, Theorem 6.6 indicates that converting this Gröbner basis G into a Gröbner basis G' with respect to the *lex* ordering would be infeasible using the FGLM algorithm, as the dimension of $R = \mathbf{F}[x_{ij}, k_{rs}]/I$ is far too large. Whether this conversion can be performed with lower complexity than the FGLM algorithm suggests, or whether it is feasible to obtain the required univariate polynomials in the key variables, are interesting research areas.

A further approach for using the Gröbner basis for the AES system would be to try to recover a key byte simply by testing whether certain solution polynomials are in the ideal generated by the AES equations [14]. A Gröbner basis provides a powerful tool to solve the ideal membership problem, as a polynomial p belongs to the ideal I if and only if p reduces to zero with respect to any Gröbner basis of I.

For the AES, the näive approach would be to verify whether the polynomial $k_{0,j} + a_j$ belongs to the ideal I generated by the AES polynomial

	Number of variables	Number of equations	Number of monomials	Time in seconds
SR(2, 1, 1, 4)	36	104	137	0.03
SR(3, 1, 1, 4)	52	152	201	0.11
SR(4, 1, 1, 4)	68	200	265	0.28
SR(5, 1, 1, 4)	84	248	339	0.97
SR(6, 1, 1, 4)	100	296	393	4.30
SR(7, 1, 1, 4)	116	344	457	11.26
SR(8, 1, 1, 4)	132	392	521	16.56
SR(9, 1, 1, 4)	148	440	585	46.05
SR(10, 1, 1, 4)	164	488	649	74.06
SR(2, 1, 1, 8)	72	172	365	118.45
SR(3, 1, 1, 8)	104	252	541	N/A

Figure 6.3. The computation time of the F4 algorithm for the system of equations generated by SR(r, 1, 1, e) over GF(2).

equations. In this case, the key byte is given by $k_{0,j} = a_j$. However, the Gröbner basis G of Section 5.4 gives solutions over the algebraic closure of **F**. Thus, even if we know that the system has unique solution in **F** given by $k_{0,j} = a_j$, all we can guarantee is that a polynomial of the form

$$q \cdot (k_{0,j} + a_j)^t$$

belongs to the ideal I associated with the AES equation system, where $q \in \mathbf{F}[k_{0,j}]$. The degree of this polynomial, as well as the exponent t, are related to the dimension of the quotient ring R. Therefore it seems unlikely that this approach can give an efficient method for solution of the AES equation system.

It is quite surprising that a Gröbner basis for the AES equation system can be obtained in such a straightforward manner [14]. Although obvious natural approaches do not seem to provide a direct solution to the key recovery problem, it is an interesting question whether the existence of such a Gröbner basis for the AES equation system can be exploited.

Experimental results

We now discuss some experimental results concerning Gröbner basis methods for solving the equation systems arising from small scale variants of the AES (Section 3.3). These results were originally presented in [22]. The experiments were performed using the computer algebra package MAGMA [77] which includes an efficient implementation of the F4 algorithm. The Gröbner bases were computed with respect to the *grevlex* monomial ordering, and the experiments were performed on a HP workstation running Windows XP with a Pentium 4 - 3GHz pro-

	Number of variables	Number of equations	Number of monomials	Time in seconds
SR(2, 1, 1, 4)	36	72	89	0.11
SR(3, 1, 1, 4)	52	104	129	0.75
SR(4, 1, 1, 4)	68	136	169	2.02
SR(5, 1, 1, 4)	84	168	209	7.47
SR(6, 1, 1, 4)	100	200	249	23.71
SR(7, 1, 1, 4)	116	232	289	56.74
SR(8, 1, 1, 4)	132	264	329	43.70
SR(9, 1, 1, 4)	148	296	369	219.38
SR(10, 1, 1, 4)	164	328	409	340.31
SR(2, 1, 1, 8)	72	144	177	43.55
SR(3, 1, 1, 8)	104	208	257	N/A

Figure 6.4. The computation time of the F4 algorithm for the system of equations generated by $SR(r, 1, 1, e)$ over $GF(2^e)$.

cessor and 1 GB of RAM. Even though these simple experiments use off-the-shelf software with limited computing resources, they are helpful as a preliminary assessment of algebraic attacks against the AES. While small scale variants might not exhibit all the features of the AES, they might provide an understanding of how various components and representations of the AES contribute to the complexity of algebraic attacks.

We first discuss the experimental results of [22] for the small scale variants $SR(r, 1, 1, 4)$ and $SR(r, 1, 1, 8)$. These small scale variants are defined in Section 3.3. Each variant gives rise to an equation system over $GF(2)$ (Section 5.2) and a BES-style equation system over $GF(2^4)$ or $GF(2^8)$ (Section 5.3). Figure 6.3 shows the experimental results for equation systems over $GF(2)$. The experimental results for the BES-style equation systems over $GF(2^4)$ or $GF(2^8)$ are given in Figure 6.4. In both Figures 6.3 and 6.4, N/A indicates that a timing is not available due to there being insufficient memory available to complete the computation. It was observed that the time to solution depended greatly on the ordering of the variables [22].

The block cipher $SR(r, 1, 1, e)$ is particularly simple and based on a 1×1 array. We would expect to easily solve the equation system of such a block cipher with many rounds. However there was insufficient memory to solve the equation system of $SR(3, 1, 1, 8)$, even though it is of comparable size to that of the easily solved equation system of $SR(6, 1, 1, 4)$. This suggests that the field relations, which are used in a different way in the BES-style equations over $GF(2^e)$, may play an important role in the computations for solving the system.

	Number of variables	Number of equations	Number of monomials	F4 time in seconds	Buchberger time in seconds
SR(1,2,1,4)	40	80	97	0.22	1.11
SR(2,2,1,4)	72	144	177	24.55	40.58
SR(3,2,1,4)	104	208	257	519.92	2649.90
SR(4,2,1,4)	136	272	337	N/A	28999.41
SR(1,2,2,4)	72	144	169	27.73	444.07
SR(2,2,2,4)	128	256	305	N/A	N/A

Figure 6.5. The computation time when using an F4 and Buchberger Gröbner basis computation for the system of equations generated by SR$(r, \cdot, \cdot, 4)$ over GF(2^4).

A comparison of the results in Figures 6.3 and 6.4 shows that timings for the equation systems of SR$(r, 1, 1, 4)$ over GF(2) are much better than those for the same block cipher over GF(2^4). However the system for SR$(2, 1, 1, 8)$ shows the opposite behaviour. Thus it is not clear whether bit–level equations generally offer a better representation than BES–style equations, particularly since MAGMA's implementation of the F4 algorithm appears to be heavily optimised for operations over GF(2). Given the highly structured and sparse nature of the BES-style equation systems over GF(2^e), we would expect a Gröbner basis algorithm that has been optimised for GF(2^e) to give the best algebraic attack against these AES variants.

We now discuss experiments on the small scale variants SR$(r, 2, 1, 4)$ and SR$(r, 2, 2, 4)$, which are block ciphers based on a 2×1 and a 2×2 array respectively [22]. Experiments to solve the equation systems of these block ciphers using the MAGMA implementations of Buchberger's algorithm and the F4 algorithm are presented in Figure 6.5, with N/A again indicating insufficient memory. Whilst we would expect Buchberger's algorithm to be slower, it should also require less memory than the F4 algorithm. We note that the equation system for SR$(4, 2, 1, 4)$ is comparable to the equation system for SR$(2, 2, 2, 4)$. However, only the latter proved to be intractable. This illustrates the important role played by the inter-word diffusion in the complexity of the computations. The diffusion of SR$(r, 2, 1, 4)$ is limited, whereas SR$(r, 2, 2, 4)$ has a similar diffusion pattern to the AES.

The block ciphers in [22] have very small key spaces and can easily be broken by exhaustive key search. However, the results in [22] are solely concerned with algebraic analysis. As such, they provide a preliminary insight into the behaviour of algebraic attacks against AES-like block ciphers, though they seem to indicate that general purpose Gröbner basis methods are unlikely to solve a full equation system arising from the AES.

Some other experiments on equation systems with a similar structure to AES equation systems, that is with layers of linear and non-linear equations repeated for many rounds, are presented in [3]. These results on very small systems seem to indicate that the maximum degree of polynomials obtained during the running of the F5 algorithm is bounded by a reasonably small value for any number of rounds. This would suggest that the complexity of solving such a system is not what we would expect from a comparable generic system. However the connection between the equation systems in [3] and the AES equation system is not sufficiently strong to conclude that an AES equation system would behave in a similar manner.

2. Linearisation Methods

Linearisation is a well-known technique for solving certain large systems of multivariate polynomial equations. Suppose we have a system of polynomial equations

$$f_1(x_1, \ldots, x_n) = 0, \ \ldots \ , f_m(x_1, \ldots, x_n) = 0$$

over a field $\mathbb{F}$, and let $\{f_1, \ldots, f_m\}$ be the associated set of polynomials in the polynomial ring $\mathbb{F}[x_1, \ldots, x_n]$. Each polynomial is a finite linear combination of monomials $x_1^{\alpha_1} x_2^{\alpha_2} \ldots x_n^{\alpha_n} = X^\alpha$ over $\mathbb{F}$, so we have

$$f_i = \sum_{\alpha \in N'} c_\alpha^i X^\alpha,$$

where $c_\alpha^i \in \mathbb{F}$ and N' is a finite subset of the set of multi-indices $\mathbb{N}^n$ (Definition 2.29). We *linearise* this system by considering the monomials X^α as new independent variables X_α to obtain a new linear system in the variables X_α. Thus to solve the system by linearisation, we can construct the matrix A_L of this resulting linear system, where A_L is given by

$$\begin{array}{c} \\ f_1 \\ \vdots \\ f_m \end{array} \begin{array}{ccc} X_\alpha & \cdots & X_{\alpha'} \\ \left(\begin{array}{ccc} c_\alpha^1 & \cdots & c_{\alpha'}^1 \\ \vdots & \ddots & \vdots \\ c_\alpha^m & \cdots & c_{\alpha'}^m \end{array} \right), \end{array}$$

and then reduce the matrix A_L to echelon form. Any solution of the original polynomial system gives a solution of this resulting linear system, so the solutions of this linear system can be checked for consistency to obtain solutions for the original polynomial equation system.

The polynomial ring $\mathbb{F}[x_1, \ldots, x_n]$ is a vector space over $\mathbb{F}$, so this method essentially computes a basis of the vector subspace generated by the polynomials $f_1, \ldots, f_m$. It is clear that the linearisation method can only be effective if the number of linearly independent polynomials in the system is approximately the same as the number of monomials. For generic systems of equations of degree d in n variables, there are about $\binom{n+d}{d}$ distinct monomials of degree at most d. For finite fields $\mathbb{F}$, we can also make use of the field relations. For example, for GF(2) we can identify x_i^2 with x_i, so there are about $N = \sum_{i=0}^{d} \binom{n}{i} \approx n^d$ distinct monomials of degree at most d, and the matrix A_L has m rows and about N columns. For this system to be soluble directly by linearisation, we generally require that $m \geq N - 1$. Thus a necessary condition for linearisation to be effective is that the polynomial system be highly overdefined.

EXAMPLE 6.7 We consider the polynomial ring $\mathbb{Q}[x, y, z]$ of polynomials in three variables over the rational numbers. Suppose we have the equation system

$$
\begin{aligned}
xyz + xz + x + 2y + z - 3 &= 0 \\
2xyz - 4xy + xz + yz - x + y &= 0 \\
xy + xz + yz + y - z - 2 &= 0 \\
2xy + y + 7 &= 0 \\
2yz + x + y + z &= 0 \\
xyz + x + 2z - 1 &= 0 \\
2xyz - xy - 3yz - 3y &= 0.
\end{aligned}
$$

This system has seven equations and seven non-constant monomials, so it is a candidate for solution by linearisation. We construct the linearisation matrix

$$
\begin{array}{cccccccc}
xyz & xy & xz & yz & x & y & z & 1 \\
\left(\begin{array}{cccccccc}
1 & 0 & 1 & 0 & 1 & 2 & 1 & -3 \\
2 & -4 & 1 & 1 & -1 & 1 & 0 & 0 \\
0 & 1 & 1 & 1 & 0 & 1 & -1 & 2 \\
0 & 2 & 0 & 0 & 0 & 1 & 0 & 7 \\
0 & 0 & 0 & 2 & 1 & 1 & 1 & 0 \\
1 & 0 & 0 & 0 & 1 & 0 & 2 & -1 \\
2 & -1 & 0 & -3 & 0 & -3 & 0 & 0
\end{array}\right).
\end{array}
$$

Applying row reduction to this matrix gives the matrix

$$
\begin{array}{cccccccc}
xyz & xy & xz & yz & x & y & z & 1 \\
\end{array}
$$

$$
\left(
\begin{array}{cccccccc}
1 & 0 & 0 & 0 & 0 & 0 & 0 & 6 \\
0 & 1 & 0 & 0 & 0 & 0 & 0 & 3 \\
0 & 0 & 1 & 0 & 0 & 0 & 0 & -6 \\
0 & 0 & 0 & 1 & 0 & 0 & 0 & 2 \\
0 & 0 & 0 & 0 & 1 & 0 & 0 & -3 \\
0 & 0 & 0 & 0 & 0 & 1 & 0 & 1 \\
0 & 0 & 0 & 0 & 0 & 0 & 1 & -2 \\
\end{array}
\right).
$$

This gives us the solution $x = 3$, $y = -1$, $z = 2$, which is the solution of the original polynomial equation system. $\qquad\square$

If we assume that the matrix A_L is a square matrix of size N, then the linearisation method has complexity of the order of N^ω field operations, where ω is the exponent of matrix reduction (Definition 2.53).

The linearisation method has been successfully used in the cryptanalysis of some LFSR-based stream ciphers [29]. However, it seems very unlikely that linearisation can be used in a straightforward manner in the analysis of block ciphers. While equation systems for the AES are overdefined, they are not sufficiently overdefined to allow us to solve them by linearisation.

The XL algorithm

The linearisation method fails when there are not enough linearly independent polynomials. Some methods have been proposed that extend the original equation system. The intention is to generate enough linearly independent equations, and then to apply the linearisation method to this new extended polynomial equation system. We now discuss the most prominent of these methods, the *extended linearisation* or *XL algorithm* [28]. The XL algorithm was specifically proposed as an efficient algorithm for algebraic attacks against certain asymmetric cryptosystems based on multivariate quadratic equation systems.

Suppose that we have a system of polynomial equations

$$
f_1(x_1, x_2, \ldots, x_n) = 0, \ \ldots \ , f_m(x_1, x_2, \ldots, x_n) = 0
$$

over a field $\mathbb{F}$ of degree at most d, and let $\{f_1, \ldots, f_m\}$ be the associated set of polynomials in the polynomial ring $\mathbb{F}[x_1, \ldots, x_n]$. The XL algorithm multiplies the polynomials in the original system by all monomials X^β up to some prescribed degree $D - d$. Thus the XL algorithm

1: **Input:** Set $F = \{f_1, \ldots, f_m\} \subset \mathbb{F}[x_1, \ldots, x_n]$ of polynomials of degree d.
2: **Output:** Set $S \subset \mathbb{F}[x_1, \ldots, x_n]$ of univariate linear equations corresponding to the solution of the system $f_j = 0$.
3:
4: $S := \emptyset$;
5: $D := d + 1$;
6: $i := 1$;
7: **repeat**
8:    Generate all products $p_{(\beta, j)} = X^\beta f_j$ for $f_j \in F$ and monomials X^β in the variables $x_i, \ldots, x_n$ of degree at most $D - d$;
9:    Consider the system consisting of equations $p_{(\alpha, j)} = 0$ and an order on the monomials such that the monomials x_i^k are the lowest. Perform Gaussian reduction on the corresponding matrix, that is solve the system by linearisation;
10:   **if** a univariate $f(x_i)$ is found **then**
11:      Solve the univariate equation to get set of solutions A_i in the algebraic closure of the field $\mathbb{F}$;
12:      Take the (unique) $a_i \in A_i$ contained in the field $\mathbb{F}$;
13:      Make $S := S \cup \{x_i - a_i\}$;
14:      Make $p_{(\alpha, j)} = p_{(\alpha, j)}(a_i) \in \mathbb{F}[x_{i+1}, \ldots, x_n]$, that is substitute $x_i = a_i$;
15:      Make $i := i + 1$;
16:   **else**
17:      Make $D := D + 1$;
18:   **end if**
19: **until** $i = n + 1$;
20:
21: **return** S

Figure 6.6.   The XL algorithm for an equation system with a unique solution.

constructs the matrix A_{XL} given by

$$
\begin{array}{c}
\\
f_1 \\
\vdots \\
X^\beta f_1 \\
\vdots \\
X^{\beta'} f_m
\end{array}
\begin{array}{c}
X_\alpha \quad \cdots \quad X_{\alpha'} \\
\left(
\begin{array}{ccc}
c_\alpha^1 & \cdots & c_{\alpha'}^1 \\
\vdots & \ddots & \vdots \\
c_{\alpha - \beta}^1 & \cdots & c_{\alpha' - \beta}^1 \\
\vdots & \ddots & \vdots \\
c_{\alpha - \beta'}^m & \cdots & c_{\alpha' - \beta'}^m
\end{array}
\right).
\end{array}
$$

The XL algorithm attempts to use linearisation on the extended equation system to find a univariate polynomial. This means that the XL algorithm can find solutions even if the matrix A_{XL} does not quite have full rank. As with linearisation, any solution of the original equation system gives a solution of the extended linear system, so solutions of the extended system can be checked for consistency to give solutions for the original system. The XL algorithm for an equation system with a unique solution in $\mathbb{F}$ is given in Figure 6.6.

The XL algorithm is supposed to be especially suitable for overdefined systems, and a number of variants of the XL algorithm have been proposed to exploit special properties of such systems in different cases [26, 30, 32]. The XL algorithm can also be modified to use the field relations $z_i^q - z_i$ of a finite field. In this case, after every multiplication, the resulting polynomials are reduced with respect to the field relations. This algorithm is called *reduced XL algorithm* in [41] and potentially reduces the size of the matrix generated by the algorithm. However, these versions of the XL algorithm have been shown to be part of the same general theoretical framework [41].

EXAMPLE 6.8 We consider the equation system over GF(23) given by

$$x^2 + 3xy + 17 = 0 \text{ and } y^2 + 7xy + 22 = 0.$$

The XL algorithm with $D = 4$ multiplies the two polynomials by the monomials $\{x, y, x^2, xy, y^2\}$ to give the following polynomials

Multiplier	$x^2 + 3xy + 17$	$y^2 + 7xy + 22$
x	$x^3 + 3x^2y + 17x$	$xy^2 + 7x^2y + 22x$
y	$x^2y + 3xy^2 + 17y$	$y^3 + 7xy^2 + 22y$
x^2	$x^4 + 3x^3y + 17x^2$	$x^2y^2 + 7x^3y + 22x^2$
xy	$x^3y + 3x^2y^2 + 17xy$	$xy^3 + 7x^2y^2 + 22xy$
y^2	$x^2y^2 + 3xy^3 + 17y^2$	$y^4 + 7xy^3 + 22y^2.$

We can now construct the corresponding matrix A_{XL}

x^4	x^3y	x^2y^2	xy^3	x^3	x^2y	xy^2	x^2	xy	x	y^4	y^3	y^2	y	1
0	0	0	0	0	0	0	1	3	0	0	0	0	0	17
0	0	0	0	1	3	0	0	0	17	0	0	0	0	0
0	0	0	0	0	1	3	0	0	0	0	0	0	17	0
1	3	0	0	0	0	0	17	0	0	0	0	0	0	0
0	1	3	0	0	0	0	0	17	0	0	0	0	0	0
0	0	1	3	0	0	0	0	0	0	0	0	17	0	0
0	0	0	0	0	0	0	0	7	0	0	0	1	0	22
0	0	0	0	0	7	1	0	0	22	0	0	0	0	0
0	0	0	0	0	0	7	0	0	0	0	1	0	22	0
0	7	1	0	0	0	0	22	0	0	0	0	0	0	0
0	0	7	1	0	0	0	0	22	0	0	0	0	0	0
0	0	0	7	0	0	0	0	0	0	1	0	22	0	0

This matrix can be reduced by row operations to give the matrix

x^4	x^3y	x^2y^2	xy^3	x^3	x^2y	xy^2	x^2	xy	x	y^4	y^3	y^2	y	1
1	0	0	0	0	0	0	0	0	0	0	0	15	0	0
0	1	0	0	0	0	0	0	0	0	0	0	4	0	2
0	0	1	0	0	0	0	0	0	0	0	0	11	0	10
0	0	0	1	0	0	0	0	0	0	0	0	2	0	12
0	0	0	0	1	0	0	0	0	0	0	17	0	2	0
0	0	0	0	0	1	0	0	0	0	0	16	0	1	0
0	0	0	0	0	0	1	0	0	0	0	10	0	13	0
0	0	0	0	0	0	0	1	0	0	0	0	16	0	1
0	0	0	0	0	0	0	0	1	0	0	0	10	0	13
0	0	0	0	0	0	0	0	0	1	0	7	0	20	0
0	0	0	0	0	0	0	0	0	0	1	0	8	0	8
0	0	0	0	0	0	0	0	0	0	0	0	0	0	0

The penultimate row gives the equation

$$y^4 + 8y^2 + 8 = (y-3)(y-11)(y-12)(y-20) = 0,$$

which gives $\{(4,3),(8,11),(15,12),(19,20)\}$ as the solution set. $\square$

Termination of the XL algorithm

Soon after it was proposed, doubts were cast on whether the XL algorithm would terminate for all inputs [81]. In fact, there are many inputs for which the XL algorithm fails to terminate. In order for the XL algorithm to terminate, the reduction of the matrix A_{XL} has to yield a univariate polynomial at every interaction of the algorithm. For the XL algorithm with parameter D, the maximum possible degree for this univariate polynomial is D. Thus if the difference between the number of columns of A_{XL} and the rank of A_{XL} is greater than D, it may be reasonable to expect that the algorithm would not yield a univariate polynomial following the reduction of A_{XL}. There exist examples where this difference always exceeds D, however large the parameter D is chosen, and the XL algorithm fails to terminate for many of these examples. That such examples exist can be demonstrated using techniques from *Hilbert Theory* [34].

The polynomial ring $\mathbb{F}[x_1, \ldots, x_n]$ is also a vector space over $\mathbb{F}$ (Example 2.40), and we let $\mathbb{F}[x_1, \ldots, x_n]_s$ and $\mathbb{F}[x_1, \ldots, x_n]_{\leq s}$ denote the subspaces generated by the monomials of degree s and the monomials of degree at most s respectively. We can now define the homomorphism

$$\nu_s \colon \mathbb{F}[x_1, \ldots, x_n]_{\leq s} \;\rightarrow\; \mathbb{F}[x_0, x_1, \ldots, x_n]_s$$
$$f \;\mapsto\; x_0^s f\left(\frac{x_1}{x_0}, \ldots \frac{x_n}{x_0}\right)$$

so $\nu_s(f)$ is a homogeneous polynomial of degree s in $n+1$ variables.

Suppose now that we have a quadratic equation system and that $\{f_1, \ldots, f_m\}$ is the associated set of polynomials in $\mathbb{F}[x_1, \ldots, x_n]$. We can define the ideal

$$\overline{I} = \langle\, \nu_2(f_1), \ldots, \nu_2(f_m) \,\rangle \lhd \mathbb{F}[x_0, x_1, \ldots, x_n],$$

and we let $\overline{I}_s = \overline{I} \cap \mathbb{F}[x_0, x_1, \ldots, x_n]_s$ be the s-homogeneous component of $\overline{I}$, that is the subset of all homogeneous polynomials of degree s in $\overline{I}$. It is shown in [41] that if

$$\chi(D) = \dim_{\mathbb{F}}(\mathbb{F}[x_0, x_1, \ldots, x_n]_D) - \dim_{\mathbb{F}}(\overline{I}_D) \leq D,$$

then the Gaussian reduction of the matrix A_{XL} in the XL algorithm with parameter D yields a non-trivial univariate polynomial. Following [41], we define D_{min} the least positive integer D such that $\chi(D) \leq D$. If there is no such D, then we say that $D_{min} = \infty$ and it is very likely that for most such systems the XL algorithm does not terminate.

The function $\chi(D)$ is called the *Hilbert Function* of $\overline{I}$ [34]. The ideal $\overline{I}$ defines a projective variety $\mathcal{W}(\overline{I})$, of which the affine portion is defined by the ideal $I = \langle f_1, \ldots, f_m \rangle$ (Section 2.5). A well-known result from Hilbert Theory states that for D large enough, the Hilbert function $\chi(D)$ is a polynomial in D, called the *Hilbert Polynomial*, with degree equal to the dimension of $\mathcal{W}(\overline{I})$ [34].

This result about the Hilbert polynomial has important consequences for the XL algorithm. If the projective variety $\mathcal{W}(\overline{I})$ has nonzero dimension, then $\chi(D)$ is a non–constant integer polynomial for large enough D. We would thus expect $\chi(D) > D$ for large enough D. In this case, the reduction of the matrix A_{XL} should not yield a univariate polynomial for most systems, and it is very likely that the XL algorithm does not terminate. However, in the equation systems that arise in cryptology, we are usually interested in the unique solution to the equation system over a small field and can therefore include the field equations $x_i^q - x_i = 0$. The associated projective variety is then zero-dimensional, that is $\dim(\mathcal{W}(\overline{I})) = 0$. This guarantees that there exists a value of D' such that the Hilbert function is constant for all $D \geq D'$, and the XL algorithm should terminate in these cases of cryptographic interest.

Complexity of the XL algorithm

We consider the complexity of solving a generic multivariate quadratic equation system with m quadratic equations and n variables using the XL algorithm. The XL algorithm attempts to solve this system by multiplying the m quadratic equations in the system by all monomials up to a prescribed degree $(D - 2)$, and then solving the resulting extended system of maximal degree D by linearisation. The number of monomials in $\mathbb{F}[x_1, \ldots, x_n]$ of degree at most D is $\binom{n+D}{D}$. Thus the XL algorithm generates an extended equation system with

$$m\binom{n+D-2}{D-2} \text{ equations in about } \binom{n+D}{D} \text{ distinct monomials.}$$

If the equation system is over GF(2), then the field relations allow us to identify x_i^2 with x_i. In this case, the (reduced) XL algorithm generates an extended equation system with

$$m\sum_{i=0}^{D-2}\binom{n}{i} \text{ equations in about } \sum_{i=0}^{D}\binom{n}{i} \text{ distinct monomials.}$$

If the extended equation system contains N monomials, then the complexity of the XL algorithm is expected to be of the order of N^ω field operations, where ω is the exponent of matrix reduction (Definition 2.53).

The key issue in obtaining lower bounds for the complexity of the XL algorithm is finding D_{min}, the minimal degree D such that $\chi(D) \leq D$. For this value of D, the XL algorithm is expected to yield a univariate polynomial. The value D is determined by the number of linearly independent polynomials (over $\mathbb{F}$) generated by each interaction of the XL algorithm. We note that it is very difficult to obtain accurate estimates for D_{min} [18, 19]. The original heuristic estimates from the XL proposal are given in [28], where it was suggested that the algorithm had subexponential complexity when $m > n$. However, these estimates proved to be too optimistic. In fact, for $m = n + c$ $(c \geq 1)$, we have

$$D_{min} \geq \frac{n}{\sqrt{c-1}+1},$$

with more precise lower bounds when $c \geq 3$ [41]. Thus the complexity of the XL algorithm does not seem to be subexponential in n.

A compact AES quadratic equation system over GF(2) has 8000 equations in 1600 variables, excluding field equations (Section 5.2). For such a system, the value $D_{min} = 18$ is suggested in [31]. This estimate is based on the original heuristic complexity estimates for the XL algorithm given in [28] and would give a complexity of the order of 2^{330} field operations to solve the AES equation system. In fact, the results of [41] show that $D_{min} \geq 44$, which gives a complexity for solving the AES equation system of at least the order of $\left(\sum_{i=0}^{44}\binom{1600}{i}\right)^\omega \approx 2^{681}$ field operations [20]. Even though variants of the XL algorithm might reduce this figure to the order of 2^{488} field operations [30], it seems that solving a generic quadratic equation system of comparable size to an AES equation system using the XL algorithm is infeasible.

Comparison between the XL and Gröbner basis algorithms

The XL algorithm was proposed specifically to solve systems of multivariate equations arising in cryptology. Such systems typically have a unique solution over a small finite field and are often overdefined. It was intended that the XL algorithm would exploit these properties to find the solution of such a system without having to compute the Gröbner basis of the associated ideal using some generic Gröbner basis algorithm. In fact the authors of the XL algorithm expected that the algorithm would be more efficient than Gröbner basis algorithms in these special cases [28]. However, there is now a much better understanding of the behaviour of the XL algorithm than when first proposed [4, 17–19, 41]. In particular, it has been shown that the XL algorithm is much more closely related to Gröbner basis algorithms than had been originally anticipated.

The relationship between the XL algorithm and Gröbner basis algorithms is considered in [4]. An analysis of a version of the XL algorithm shows that the XL algorithm works in a similar manner to the F4 calculation of a Gröbner basis of a polynomial ideal with the *lex* ordering. In the case of an equation system over a finite field with a unique solution, the XL algorithm computes the Gröbner basis of the associated ideal of the equation system and the field equations and it can be seen as a redundant version of the F4 algorithm. For such an equation system over $GF(2)$, the degree D for the XL algorithm should be roughly the same as the degree of the polynomials required by the F5 algorithm. While the matrix used by the F5 algorithm is expected to be smaller than the matrix used by the XL algorithm, for an equation system over $GF(q)$ with $q \gg n$, it is unlikely that reductions by the field relations occur. In such cases, the degree D required by the XL algorithm is likely to be higher than that required by the F5 algorithm using *grevlex* ordering [4].

The results of [4] therefore seem to indicate that the XL algorithm in its current form offers little benefit over efficient versions of generic Gröbner basis algorithms such as F4 and F5.

3. Specialised Methods

The discussions of previous sections seem to indicate that there is little hope that general techniques from computer algebra might be used in a straightforward manner to solve the equation systems arising from modern block ciphers. However, equation systems arising from iterated block ciphers can be viewed as iterated systems of equations, with similar blocks of multivariate quadratic equations repeated for every round. These blocks are connected to each other by their input and output variables and by the key schedule. These equation systems are highly sparse.

Thus it might be more promising to apply some dedicated method, perhaps built on known techniques from computer algebra, but aiming to exploit the special properties of a target system. We discuss some of these proposals below.

The XSL Algorithm

The *extended sparse linearisation* or *XSL algorithm* was designed to exploit the structure of equation systems arising from iterated block ciphers. The XSL algorithm was introduced in [31, 32], where its proposed application to the AES equation system attracted much attention.

The XSL algorithm is based on the XL algorithm (Section 6.2). In the XSL algorithm the equations are multiplied only by "carefully selected monomials", whereas the equations are multiplied by all monomials up to a certain degree in the XL algorithm. This is the core idea in the XSL algorithm and is intended to generate a large number of equations whose terms are the product of selected monomials. The goal is therefore to create fewer new monomials while generating many new equations in the extended equation system. The XSL algorithm also incorporates an additional last step called the T' *method*, in which new linearly independent equations are generated without creating any new monomials.

Different versions of the XSL algorithm have been published. The first version was proposed in [31], where two different attacks on the AES based on the XSL algorithm are described. The first version requires a few plaintext–ciphertext pairs in order to eliminate the key variables and key schedule equations as a preliminary step. The second version should require only a single plaintext–ciphertext pair and uses the key schedule equations. The *compact XSL* algorithm is a slightly different version of the algorithm and was introduced later [32]. A heuristic description of the steps in an XSL-type algorithm is given in Figure 6.7

The basic idea behind the XSL algorithm is to expand the original system by multiplying equations only by the product of monomials that already exist in the original equation system. For a sparse equation system, this potentially decreases the number of monomials generated in the extended equation system when compared to the extended system generated by the XL algorithm. Furthermore, as the XSL algorithm is based on the linearisation method, the algorithm should benefit from overdefined systems.

The XSL algorithm was intended to exploit the structure of some types of block cipher. In its basic version, the XSL algorithm assumes that the block cipher is built with layers of small S-Boxes interconnected by a key-dependent affine transformation. It is further assumed that the S-box can be described by an overdefined set of quadratic equations. For

1: **Input:** System of block cipher polynomial equations $f_1 = \ldots = f_m = 0$.
2: **Output:** Solution $(a_1, \ldots, a_n)$ where $f_1(a_1, \ldots, a_n) = \ldots = f(a_1, \ldots, a_n) = 0$.
3:
4: Choose certain sets of monomials and equations which are to be used during the
   later steps of the algorithm by analysing the original equation set.
5: Select the value of the parameter P and multiply the chosen equations by the
   product of $P - 1$ selected monomials.
6: Perform the T′ method in which some selected equations are multiplied by single
   variables.
7: Iterate T′ with as many variables as necessary until the extended system has
   enough linearly independent equations to apply linearisation.
8:
9: **Return** Solution of the extended system obtained by linearisation

Figure 6.7. Heuristic description of an XSL-type algorithm.

example, the AES and the block cipher SERPENT [9] both use S-boxes
that give rise to an overdefined systems of quadratic equations. The
equation systems for such block ciphers are sparse and XSL is supposed
to take advantage of this when expanding the system prior to linearisation. For versions of the XSL algorithm that use key schedule equations,
the key schedule should have a similar structure to the state encryption
transformation. This is the case for the AES.

XSL analysis

The XSL algorithm could be considered a general method for solving
certain structured but sparse systems of quadratic equations. However,
it was the proposal to apply XSL to the equation system of the AES
that attracted much attention.

Estimates for the complexity of the XSL algorithm are given in [32]
and refer to an analysis without the key schedule. These estimates give
a complexity of the order of 2^{298} field operations to solve the equation
system arising from the AES with 256-bit keys. It was also claimed that
much better results could be obtained when applying the complexity
estimates to equation systems including the key schedule, especially if
the BES-style equation system for the AES is used (Section 5.3). In
such a case, the complexity estimates for solving an equation system
over $GF(2^8)$ for the AES with 128-bit keys suggested that around 2^{108}
field operations would be required, which is roughly equivalent to 2^{100}
AES encryptions. This implied that if the XSL algorithm complexity
estimates of [32] were correct, then AES key recovery might have been
possible with a lower work effort than exhaustive key search [89, 90].

However, the heuristic estimates for the complexity of the XSL algorithm in [32] are too optimistic and overestimate the number of linearly independent equations generated by the algorithm. This is shown by the application of the XSL complexity estimates to the BES-style equations for the AES given by [90]. There it is demonstrated that such estimates would give significantly more linearly independent equations than terms, even though this is clearly impossible.

The heuristic XSL complexity estimates of [32] overestimate the number of linearly independent equations for two reasons. Firstly, it is assumed that all equations generated by the method are linearly independent. This is clearly not the case. For example, if f_i and f_j are two polynomials in the initial quadratic system defined as

$$f_i = \sum_{\alpha \in N_i'} c_\alpha^i X^\alpha \text{ and } f_j = \sum_{\beta \in N_j'} c_\beta^j X^\beta,$$

then even for $P = 2$ we have the relation

$$f_i \cdot [f_j] = \sum_{\alpha \in N_i'} c_\alpha^i X^\alpha \cdot f_j = \sum_{\beta \in N_j'} c_\beta^j X^\beta \cdot f_i = f_j \cdot [f_i].$$

There are many relations of this type. Secondly, the XSL algorithm states that neighbouring S-Boxes need to be excluded when multiplying the linear equations, but this is not taken into account in the estimates. Revised estimates for the XSL complexity are given in [20].

Whilst doubts were very quickly cast on the general idea of the XSL algorithm, until recently very little was known about its detailed behaviour. There are a number of reasons for this. Firstly, the XSL algorithm as proposed in [31, 32] relies on the system having a special form, and this makes it harder to give a precise mathematical analysis of the algorithm. Secondly, different versions of the XSL algorithm were published, and in all cases the description left room for interpretation. Furthermore, given the size of the systems involved, it is very difficult to implement and run experiments even on small examples to examine the heuristic arguments of [31, 32].

A detailed analysis of the XSL algorithm is presented in [20], including a simulation on a small variant of the AES. It is shown that the XSL algorithm of [32] cannot solve the equation system arising from the AES. The problem arises from the way the original equations are processed prior to multiplication and the selection of monomials. An alternative to the XSL algorithm is also discussed, but generally it appears unlikely that analytical techniques of this type would be successful [20].

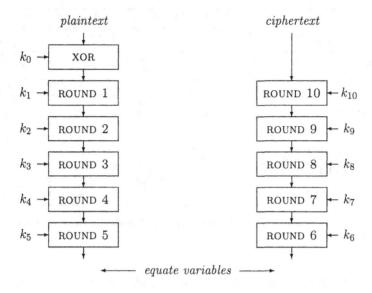

Figure 6.8. An illustration of the meet–in–the–middle technique for the AES.

Meet–in–the–middle techniques

The iterative nature of block ciphers means that the associated equation systems are typically structured in blocks. For example, the equation system for the AES consists of ten similar blocks of multivariate quadratic equations, each block containing the equations for one round. Variables in one block only occur in neighbouring blocks or within the relevant part of the key schedule.

A promising technique to find the overall solution for such equation systems is to employ a *meet-in-the-middle* approach. We divide the equation system for r rounds into two subsystems for $\frac{r}{2}$ rounds, where we assume without loss of generality that r is even. We regard the output variables of the first equation subsystem as the input variables of the second equation subsystem. We can then make use of Theorem 6.5 to simplify the problem. Theorem 6.5 gives a method for obtaining all the relations between variables in a subset S' of the set of variables S. We first compute the Gröbner basis G with respect to the *lex* ordering of the ideal generated by the set of polynomials associated with the equation system. We then extract those polynomials in the Gröbner basis G which involve only those variables in the subset S'. This technique essentially eliminates all variables in the set $S \setminus S'$.

	Number of variables	Number of equations	Number of monomials	Time in seconds
SR(10,1,1,4) five rounds →	88	172	217	19.22
SR(10,1,1,4) five rounds ←	76	148	189	22.41
Solution	16	40	52	0.02
Total				41.65
SR(4,1,1,8) two rounds →	80	152	193	15466.37
SR(4,1,1,8) two rounds ←	56	104	137	4603.89
Solution	32	80	576	215.92
Total				20286.18
SR(4,2,1,4) two rounds →	80	152	193	667.17
SR(4,2,1,4) two rounds ←	56	104	137	2722.43
Solution	80	176	524	14.87
Total				3404.47

Figure 6.9. Experimental results on the meet–in–the–middle approach when using F4 with the *lex* ordering.

For the AES equation system with r rounds, we compute the Gröbner bases of the two equation subsystems with respect to the *lex* ordering. We then eliminate variables that do not appear in rounds $\frac{r}{2}$ and $\frac{r}{2} + 1$. This gives two small systems of equations in variables from the two systems that are simply related by the round keys. These two equation systems can then be combined with some additional equations from the key schedule and solved to obtain the key (Figure 6.8).

Experimental results using this approach on AES variants are given in [22] and presented in Figure 6.9. These results seem to confirm that a meet–in–the–middle technique may be more efficient than solving the full AES equation system directly. For example, the equation system for the small scale AES variant $SR(10, 1, 1, 4)$ (Section 3.3) can be solved with the meet–in–the–middle approach in 42 seconds, whereas the direct approach took 340 seconds (Figure 6.4). Better results were also obtained for $SR(4, 1, 1, 8)$ and $SR(4, 2, 1, 4)$ using the meet–in–the–middle approach.

The meet–in–the–middle approach is cryptographically intuitive and can be considered within the context of time-memory trade-off attacks. In a chosen-plaintext attack, the first subsystem of equations does not change and solving this equation subsystem can be considered precomputation. Although the exact complexity and storage requirements for this precomputation phase are not clear, such precomputation would clearly reduce the time complexity of the online attack.

One possible drawback to this approach is that computations using elimination orderings such as *lex* are often less efficient than those with

	Number of variables	Number of equations	Number of monomials	Time in seconds
SR(4,2,1,4) two rounds →	112	216	273	553.63
SR(4,2,1,4) two rounds ←	104	200	257	1501.41
Solution	136	1197	918	12.68
			Total	2067.72

Figure 6.10. Experimental results on the meet–in–the–middle approach when using F4 with the *grevlex* ordering.

degree orderings such as *grevlex*. Thus we might expect that using the *lex* ordering in the two subsystems would give only limited advantages over using the *grevlex* ordering for the full system. An alternative approach would be to simply compute the Gröbner bases for the two subsystems using the most efficient ordering and then to combine both results to compute the solution of the full set equations. Some experimental results on this approach from [22] are given in Figure 6.10. These indicate that this approach was more efficient for the small scale variant $SR(4, 2, 1, 4)$, though more expensive for the small scale variant $SR(10, 1, 1, 4)$.

In general, however, the experimental results of [22] suggest the applicability of a more general *divide–and–conquer* approach to the problem of solving the equation system deriving from the AES. In such an approach, some form of (perhaps largely symbolic) pre-computation could take place. This could then be used to produce the solution to the full equation system.

Chapter 7

CLOSING REMARKS

In this monograph we have provided a summary of topics in the algebraic analysis of the AES. While appealing to algebraic techniques in cryptanalysis is not new, the range of algebraic techniques that might be used in an analysis of the AES is unprecedented. No other major block cipher offers quite the same opportunities. The AES has a round function that can be described in a particularly succinct fashion and the entire encryption process can be represented by a compact and simple system of equations.

The designers of the AES were careful to ensure that traditional techniques would not be useful in any cryptanalysis. Thus techniques from areas of mathematics and computer science not ordinarily considered in cryptology may be needed for any substantive cryptanalysis of the AES. The most direct method of cryptanalysis would be to recover an AES key by solving an equation system. However, it is unlikely that any general purpose method for doing this would be successful against the AES. Instead, it is almost certain that techniques tailored to the AES would be required.

We anticipate that a greater understanding of the algebraic properties of the AES will be developed in the coming years. However, there are currently no techniques, algebraic or otherwise, that compromise the security of the AES.

Appendix A
Inversion Equations over $\mathbf{GF}(2)$

We assume the bit ordering $(w_7, \ldots, w_0)^T$ for bytes in the AES. The matrices T_θ and S, corresponding respectively to multiplying by θ and squaring in $\mathbf{F}$, are given in Example 2.65. As described in Section 5.2, for $w, x \neq 0$ we have

$$C_w x = (0, 0, 0, 0, 0, 0, 0, 1)^T$$

where $C_w x$ is given by

$$
\begin{pmatrix}
\begin{aligned}
&w_7 x_7 + w_7 x_5 + w_7 x_4 + w_7 x_0 + w_6 x_6 + w_6 x_5 + w_6 x_1 + w_5 x_7 \\
&+ w_5 x_6 + w_5 x_2 + w_4 x_7 + w_4 x_3 + w_3 x_4 + w_2 x_5 + w_1 x_6 + w_0 x_7
\end{aligned} \\
\hline
\begin{aligned}
&w_7 x_6 + w_7 x_4 + w_7 x_3 + w_6 x_7 + w_6 x_5 + w_6 x_4 + w_6 x_0 + w_5 x_6 + w_5 x_5 \\
&+ w_5 x_1 + w_4 x_7 + w_4 x_6 + w_4 x_2 + w_3 x_7 + w_3 x_3 + w_2 x_4 + w_1 x_5 + w_0 x_6
\end{aligned} \\
\hline
\begin{aligned}
&w_7 x_5 + w_7 x_3 + w_7 x_2 + w_6 x_6 + w_6 x_4 + w_6 x_3 \\
&+ w_5 x_7 + w_5 x_5 + w_5 x_4 + w_5 x_0 + w_4 x_6 + w_4 x_5 + w_4 x_1 \\
&+ w_3 x_7 + w_3 x_6 + w_3 x_2 + w_2 x_7 + w_2 x_3 + w_1 x_4 + w_0 x_5
\end{aligned} \\
\hline
\begin{aligned}
&w_7 x_7 + w_7 x_4 + w_7 x_2 + w_7 x_1 + w_6 x_5 + w_6 x_3 + w_6 x_2 \\
&+ w_5 x_6 + w_5 x_4 + w_5 x_3 + w_4 x_7 + w_4 x_5 + w_4 x_4 + w_4 x_0 + w_3 x_6 \\
&+ w_3 x_5 + w_3 x_1 + w_2 x_7 + w_2 x_6 + w_2 x_2 + w_1 x_7 + w_1 x_3 + w_0 x_4
\end{aligned} \\
\hline
\begin{aligned}
&w_7 x_7 + w_7 x_6 + w_7 x_5 + w_7 x_4 + w_7 x_3 + w_7 x_1 + w_6 x_7 + w_6 x_6 \\
&+ w_6 x_5 + w_6 x_4 + w_6 x_2 + w_5 x_7 + w_5 x_6 + w_5 x_5 + w_5 x_3 + w_4 x_7 + w_4 x_6 \\
&+ w_4 x_4 + w_3 x_7 + w_3 x_5 + w_3 x_0 + w_2 x_6 + w_2 x_1 + w_1 x_7 + w_1 x_2 + w_0 x_3
\end{aligned} \\
\hline
\begin{aligned}
&w_7 x_6 + w_7 x_3 + w_7 x_2 + w_6 x_7 + w_6 x_4 + w_6 x_3 + w_5 x_5 + w_5 x_4 \\
&+ w_4 x_6 + w_4 x_5 + w_3 x_7 + w_3 x_6 + w_2 x_7 + w_2 x_0 + w_1 x_1 + w_0 x_2
\end{aligned} \\
\hline
\begin{aligned}
&w_7 x_7 + w_7 x_5 + w_7 x_2 + w_7 x_1 + w_6 x_6 + w_6 x_3 + w_6 x_2 \\
&+ w_5 x_7 + w_5 x_4 + w_5 x_3 + w_4 x_5 + w_4 x_4 + w_3 x_6 \\
&+ w_3 x_5 + w_2 x_7 + w_2 x_6 + w_1 x_7 + w_1 x_0 + w_0 x_1
\end{aligned} \\
\hline
\begin{aligned}
&w_7 x_6 + w_7 x_5 + w_7 x_1 + w_6 x_7 + w_6 x_6 + w_6 x_2 \\
&+ w_5 x_7 + w_5 x_3 + w_4 x_4 + w_3 x_5 + w_2 x_6 + w_1 x_7 + w_0 x_0
\end{aligned}
\end{pmatrix}
$$

The above matrix equation gives seven multivariate quadratic equations over $\mathrm{GF}(2)$ for AES inversion. A further equation is given with high probability.

As also described in Section 5.2, we have

$$(C_w S + I)x = 0$$

where the vector $(C_w S + I)x$ is given by

$$
\begin{pmatrix}
\begin{array}{l}
w_7 x_6 + w_7 x_5 + w_7 x_2 + w_7 x_0 + w_6 x_7 + w_6 x_4 + w_6 x_3 + w_5 x_7 + w_5 x_6 \\
+ w_5 x_3 + w_5 x_1 + w_4 x_5 + w_4 x_4 + w_3 x_7 + w_3 x_4 + w_3 x_2 + w_2 x_6 + w_2 x_5 \\
\qquad\qquad + w_1 x_5 + w_1 x_3 + w_0 x_7 + w_0 x_6 + x_7 \\
\hline
w_7 x_6 + w_7 x_3 + w_7 x_2 + w_6 x_6 + w_6 x_5 + w_6 x_2 + w_6 x_0 + w_5 x_7 + w_5 x_4 \\
+ w_5 x_3 + w_4 x_7 + w_4 x_6 + w_4 x_3 + w_4 x_1 + w_3 x_5 + w_3 x_4 + w_2 x_7 + w_2 x_4 \\
\qquad\qquad + w_2 x_2 + w_1 x_6 + w_1 x_5 + w_0 x_5 + w_0 x_3 + x_6 \\
\hline
w_7 x_7 + w_7 x_5 + w_7 x_4 + w_7 x_1 + w_6 x_6 + w_6 x_3 + w_6 x_2 + w_5 x_6 + w_5 x_5 \\
+ w_5 x_2 + w_5 x_0 + w_4 x_7 + w_4 x_4 + w_4 x_3 + w_3 x_7 + w_3 x_6 + w_3 x_3 + w_3 x_1 \\
\qquad\qquad + w_2 x_5 + w_2 x_4 + w_1 x_7 + w_1 x_4 + w_1 x_2 + w_0 x_6 + w_0 x_5 + x_5 \\
\hline
w_7 x_7 + w_7 x_5 + w_7 x_2 + w_7 x_1 + w_6 x_7 + w_6 x_5 + w_6 x_4 + w_6 x_1 + w_5 x_6 \\
+ w_5 x_3 + w_5 x_2 + w_4 x_6 + w_4 x_5 + w_4 x_2 + w_4 x_0 + w_3 x_7 + w_3 x_4 + w_3 x_3 \\
+ w_2 x_7 + w_2 x_6 + w_2 x_3 + w_2 x_1 + w_1 x_5 + w_1 x_4 + w_0 x_7 + w_0 x_4 + w_0 x_2 \\
\qquad\qquad + x_4 \\
\hline
w_7 x_5 + w_7 x_4 + w_7 x_3 + w_7 x_2 + w_6 x_5 + w_6 x_4 + w_6 x_3 + w_6 x_2 + w_6 x_1 \\
+ w_5 x_6 + w_5 x_5 + w_5 x_4 + w_5 x_3 + w_4 x_6 + w_4 x_5 + w_4 x_4 + w_4 x_3 + w_4 x_2 \\
+ w_3 x_7 + w_3 x_6 + w_3 x_5 + w_3 x_4 + w_3 x_0 + w_2 x_7 + w_2 x_6 + w_2 x_5 + w_2 x_4 \\
+ w_2 x_3 + w_1 x_7 + w_1 x_6 + w_1 x_5 + w_1 x_1 + w_0 x_7 + w_0 x_6 + w_0 x_5 + w_0 x_4 \\
\qquad\qquad + x_3 \\
\hline
w_7 x_7 + w_7 x_6 + w_7 x_5 + w_7 x_4 + w_7 x_3 + w_7 x_1 + w_6 x_7 + w_6 x_5 + w_6 x_2 \\
+ w_5 x_7 + w_5 x_6 + w_5 x_5 + w_5 x_4 + w_5 x_2 + w_4 x_6 + w_4 x_3 + w_3 x_7 + w_3 x_6 \\
+ w_3 x_5 + w_3 x_3 + w_2 x_7 + w_2 x_4 + w_2 x_0 + w_1 x_7 + w_1 x_6 + w_1 x_4 + w_0 x_5 \\
\qquad\qquad + w_0 x_1 + x_2 \\
\hline
w_7 x_6 + w_7 x_4 + w_7 x_1 + w_6 x_7 + w_6 x_6 + w_6 x_5 + w_6 x_4 + w_6 x_3 + w_6 x_1 \\
+ w_5 x_7 + w_5 x_5 + w_5 x_2 + w_4 x_7 + w_4 x_6 + w_4 x_5 + w_4 x_4 + w_4 x_2 + w_3 x_6 \\
+ w_3 x_3 + w_2 x_7 + w_2 x_6 + w_2 x_5 + w_2 x_3 + w_1 x_7 + w_1 x_4 + w_1 x_0 + w_0 x_7 \\
\qquad\qquad + w_0 x_6 + w_0 x_4 + x_1 \\
\hline
w_7 x_7 + w_7 x_4 + w_7 x_3 + w_6 x_7 + w_6 x_6 + w_6 x_3 + w_6 x_1 + w_5 x_5 + w_5 x_4 \\
+ w_4 x_7 + w_4 x_4 + w_4 x_2 + w_3 x_6 + w_3 x_5 + w_2 x_5 + w_2 x_3 + w_1 x_7 + w_1 x_6 \\
\qquad\qquad + w_0 x_6 + w_0 x_4 + w_0 x_0 + x_0
\end{array}
\end{pmatrix}
$$

This gives eight more multivariate quadratic equations over GF(2) for AES inversion. We also have

$$(C_x S + I)w = 0$$

where $(C_x S + I)x$ is given by the above vector with w and x interchanged. This gives eight further multivariate quadratic equations over GF(2) for AES inversion.

Appendix B
Augmented Linear Diffusion Matrices

The augmented linear diffusion matrix M with respect to the standard basis is given on the next two pages. We use . to represent 0 in this matrix, and the 128-bit inputs to the augmented linear diffusion layer are viewed as column vectors. We note that M is a block matrix built from three different nonzero submatrices, which correspond to the transformations consisting of the GF(2)-linear map followed by multiplication by 1, θ or $\theta + 1$ (01, 02, 03 in hexadecimal) respectively.

We then give the augmented linear diffusion matrix $P^{-1}MP$, where the matrix P is given in [88]. This is just the augmented linear diffusion matrix M with respect to a different basis. We again use . to represent 0 in this matrix, which acts on column vectors. The matrix $P^{-1}MP$ is a block diagonal matrix, and line breaks in the presentation of this matrix represent the division into fifteen invariant subspaces.

Augmented linear diffusion matrix M

```
.11111.................................1....1....................................11111.................................11111...
...11111...............................1....1....................................11111.................................11111...
....11111..............................1....1....................................11111.................................11111.
..111..111.............................11.1.......................................11111.................................11111
...111111..............................1.11.....................................1...11111................................1...11111
111...11...............................1....1.....................................11...111................................11...111
....1..1.1............................111.1.1....................................111...11................................111...11
11111..................................1....1....................................1111...1................................1111...1
11111..............................11111........................................1....1...................................11111...
.11111.............................11111.........................................1....1...................................11111.
..11111............................11111.........................................1....1...................................11111.
...11111...........................111.111.......................................11.1.....................................11111
1...1111...........................111111........................................1.11.....................................1...1111
11...11...........................111...11........................................1....1...................................11...111
111...11...........................1....1........................................111.1.1..................................111...11
1111...1...........................11111.........................................1....1...................................1111...1
11111..............................11111.........................................11111.....................................1...1.
.11111.............................11111.........................................11111.....................................1....1.
..11111............................11111.........................................11111.....................................1....1
...11111...........................11111........................................111.111...................................11.1...
1...1111...........................1...1111......................................111111....................................1.11..
11...11............................11...111.......................................111...11..................................1....1.
111...11...........................111...11.......................................1....1...................................111.1.1.
1111...1...........................1111...1.......................................11111.....................................1...1
1....1.............................11111.........................................11111......................................11111.
.1....1............................11111.........................................11111......................................11111.
..1....1...........................11111.........................................11111......................................11111
.11.1..............................11111.........................................11111.....................................111..111
1.11...............................1...1111......................................1...1111...................................111111
..1...1............................11...111.......................................11...111..................................111...11
111.1.1............................111...11.......................................111...11....................................1...1
..1...1............................1111...1.......................................1111...1...................................11111..
.....................11111....11111...................................1....1....................................11111........
.....................11111....11111...................................1....1....................................11111........
.....................11111....11111...................................1....1....................................11111........
.....................11111.111.111....................................11.1.......................................11111........
.....................1....1111..111111................................1.11.......................................1....1111......
.....................11...111111...11.................................1....1......................................11...111......
.....................111...11....1..1.................................111.1.1.....................................111...11......
.....................11111...11111....................................1....1......................................1111...1......
.....................11111...11111....................................11111.......................................1....1.....
.....................11111...11111....................................11111.......................................1....1.....
.....................11111...11111....................................11.111.......................................11.1.....
.....................1...11111...1111.................................111111.......................................1.11.....
.....................11...11111...1111................................111...11......................................1....1.....
.....................111...11111...11.................................1....1.......................................111.1.1.....
.....................1111...11111...1.................................11111.........................................1...1.....
.....................1....1..11111....................................11111.........................................11111.....
.....................1....1....11111..................................11111.........................................11111.....
.....................1....1...11111...................................11111.........................................11111.....
.....................11.1........11111................................11.111........................................111..111.....
.....................1.11...1....1111.................................1...1111.......................................111111.....
.....................1..1..11...111...................................11...111.......................................111...11.....
.....................111.1.1.111...11.................................111...11........................................1...1.....
.....................1..11111....1....................................1111...1........................................11111.....
.....................11111..1....1....................................11111..........................................11111.....
.....................11111..1....1....................................11111..........................................11111.....
.....................111.111.11.1.....................................11111..........................................11111.....
.....................1111111.11.......................................1...1111.......................................1...1111.....
.....................111...11...1.....................................11...111.......................................11...111.....
.....................1..1111.1.1......................................111...11.......................................111...11.....
.....................11111.......1..1.................................1111...1........................................1111...1.....
```

```
...............11111....................................11111....11111...................................1....1........
...............11111.....................................11111....11111...................................1....1.......
................11111....................................11111....11111...................................1....1.......
................11111.....................................11111.111.111...................................11.1........
..............1...1111....................................1...1111..111111.................................1.11........
...............11...111...................................11...111111...11..................................1..1.......
...............111...11...................................111...11....1..1.................................111.1.1.....
...............1111...1....................................1111...111111....................................1...1......
...............1....1......................................11111....11111....................................11111......
...............1....1......................................11111....11111....................................11111......
..............1....1.......................................11111....11111....................................11111......
...............11.1........................................11111....11111...................................111.111.....
...............1.11........................................1...11111...1111.................................1111111.....
..............1....1........................................11...11111...111..................................111...11....
...............111.1.1......................................111....1...11..................................1....1......
..............1....1........................................1111....11111...1..............................11111......
...............11111........................................1....1...11111..................................11111......
...............11111.........................................1....1..11111...................................11111......
...............11111.........................................1....1..11111...................................11111......
..............111.111.......................................11.1.....11111...................................11111......
...............111111........................................1.11....1...1111...............................1...1111.....
...............111...11......................................1....1..11...111................................11...111....
..............1....1........................................111.1.1.111...11................................111...11....
...............11111........................................1....11111...1...................................1111...1....
...............11111.........................................11111...1....1..................................11111......
...............11111.........................................11111...1....1..................................11111......
...............11111........................................111.111.11.1.....................................11111......
..............1...1111......................................1111111.11.......................................1....1111...
...............11...111......................................111....11...1..1...............................11...111....
...............111...11......................................1...1111.1.1....................................111...11....
...............1111...1......................................11111......1...1................................1111...1....
..............1....1.........................11111....................................11111....11111............
...............1....1.........................11111....................................11111....11111...........
..............1....1.........................11111....................................11111....11111...........
...............11.1..........................11111...................................11111.111.111...........
...............1.11...........................1...1111................................1...1111..111111........
..............1....1.........................11...111.................................11...111111...11.......
...............111.1.1.......................111....11................................111...11....1..1.......
..............1....1.........................1111...1..................................1111...111111........
...............11111.........................1....1...................................11111....11111........
...............11111..........................1....1...................................11111....11111.......
...............11111..........................1....1...................................11111....11111.......
...............111.111........................11.1.....................................11111....11111.......
...............111111..........................1.11...................................1...11111...1111.....
...............111...11.......................1....1....................................11...11111...111....
..............1....1..........................111.1.1..................................111....11111...11....
...............11111..........................1....1...................................1111...11111...1....
...............11111.........................11111....................................1....1...11111.......
...............11111..........................11111....................................1....1..11111.......
...............11111..........................11111....................................1....1..11111.......
..............1...1111........................111.111................................11.1.....11111.......
...............11...111.......................11111....................................1.11....1...1111.....
..............111...11.........................111...11.................................1....1..11...111....
...............1111...1.......................1....1..................................111.1.1.111...11......
...............11111...1......................11111...................................1....11111...1.......
...............11111..........................11111...................................11111...1....1......
...............11111...........................11111..................................11111...1....1.......
...............11111...........................11111.................................111.111.11.1.........
..............1....1111.......................1....1111...............................1111111.11..........
...............11...111.......................11....111...............................111....11...1..1.....
..............111...11.........................111....11................................1..1111.1.1........
...............1111...1........................1111....1...............................11111......1...1....
```

Augmented linear diffusion matrix $P^{-1}MP$

Appendix C
Equation System for $SR(2,2,2,4)$ over $GF(2^4)$

We illustrate the style of the equation system over $GF(2^8)$ for the AES by giving the equation system over $GF(2^4)$ for the small scale variant $SR(2,2,2,4)$ of the AES (Section 3.3 and [22]). These are BES-style equations for which we assume that 0-inversions do not occur.

Component j and conjugate l for the plaintext, ciphertext and the key (also used as the initial round key) are denoted by p_{jl}, c_{jl} and k_{0jl} respectively. We regard the two rounds as round one and round two. We denote the input and output of the inversion and the subkey used in round i for component j and conjugate l by w_{ijl}, x_{ijl} and k_{ijl} respectively.

	$GF(2^4)$ Variable	Round i	Component j	Conjugate l
Input/Output				
Plaintext	p_{jl}		$0,1,2,3$	$0,1,2,3$
Ciphertext	c_{jl}		$0,1,2,3$	$0,1,2,3$
State				
Inversion input	w_{ijl}	$1,2$	$0,1,2,3$	$0,1,2,3$
Inversion output	x_{ijl}	$1,2$	$0,1,2,3$	$0,1,2,3$
Key				
Subkey	k_{ijl}	$0,1,2$	$0,1,2,3$	$0,1,2,3$
Dummy	s_{ijl}	$0,1$	$0,1,2,3$	$0,1,2,3$

Initial Subkey Relations

$$
\begin{array}{llll}
w_{100} + p_{00} + k_{000} & w_{110} + p_{10} + k_{010} & w_{120} + p_{20} + k_{020} & w_{130} + p_{30} + k_{030} \\
w_{101} + p_{01} + k_{001} & w_{111} + p_{11} + k_{011} & w_{121} + p_{21} + k_{021} & w_{131} + p_{31} + k_{031} \\
w_{102} + p_{02} + k_{002} & w_{112} + p_{12} + k_{012} & w_{122} + p_{22} + k_{022} & w_{132} + p_{32} + k_{032} \\
w_{103} + p_{03} + k_{003} & w_{113} + p_{13} + k_{013} & w_{123} + p_{23} + k_{023} & w_{133} + p_{33} + k_{033}
\end{array}
$$

Inversion and Conjugacy Relations: Rounds 1 and 2

$$
\begin{array}{lll}
w_{100}^2 + w_{101} & w_{100}x_{100} + 1 & x_{100}^2 + x_{101} \\
w_{200}^2 + w_{201} & w_{200}x_{200} + 1 & x_{200}^2 + x_{201} \\
w_{101}^2 + w_{102} & w_{101}x_{101} + 1 & x_{101}^2 + x_{102} \\
w_{201}^2 + w_{202} & w_{201}x_{201} + 1 & x_{201}^2 + x_{202} \\
w_{102}^2 + w_{103} & w_{102}x_{102} + 1 & x_{102}^2 + x_{103} \\
w_{202}^2 + w_{203} & w_{202}x_{202} + 1 & x_{202}^2 + x_{203} \\
w_{103}^2 + w_{100} & w_{103}x_{103} + 1 & x_{103}^2 + x_{100} \\
w_{203}^2 + w_{200} & w_{203}x_{203} + 1 & x_{203}^2 + x_{200} \\
w_{110}^2 + w_{111} & w_{110}x_{110} + 1 & x_{110}^2 + x_{111} \\
w_{210}^2 + w_{211} & w_{210}x_{210} + 1 & x_{210}^2 + x_{211} \\
w_{111}^2 + w_{112} & w_{111}x_{111} + 1 & x_{111}^2 + x_{112} \\
w_{211}^2 + w_{212} & w_{211}x_{211} + 1 & x_{211}^2 + x_{212} \\
w_{112}^2 + w_{113} & w_{112}x_{112} + 1 & x_{112}^2 + x_{113} \\
w_{212}^2 + w_{213} & w_{212}x_{212} + 1 & x_{212}^2 + x_{213} \\
w_{113}^2 + w_{110} & w_{113}x_{113} + 1 & x_{113}^2 + x_{110} \\
w_{213}^2 + w_{210} & w_{213}x_{213} + 1 & x_{213}^2 + x_{210} \\
w_{120}^2 + w_{121} & w_{120}x_{120} + 1 & x_{120}^2 + x_{121} \\
w_{220}^2 + w_{221} & w_{220}x_{220} + 1 & x_{220}^2 + x_{221} \\
w_{121}^2 + w_{122} & w_{121}x_{121} + 1 & x_{121}^2 + x_{122} \\
w_{221}^2 + w_{222} & w_{221}x_{221} + 1 & x_{221}^2 + x_{222} \\
w_{122}^2 + w_{123} & w_{122}x_{122} + 1 & x_{122}^2 + x_{123} \\
w_{222}^2 + w_{223} & w_{222}x_{222} + 1 & x_{222}^2 + x_{223} \\
w_{123}^2 + w_{120} & w_{123}x_{123} + 1 & x_{123}^2 + x_{120} \\
w_{223}^2 + w_{220} & w_{223}x_{223} + 1 & x_{223}^2 + x_{220} \\
w_{130}^2 + w_{131} & w_{130}x_{130} + 1 & x_{130}^2 + x_{131} \\
w_{230}^2 + w_{231} & w_{230}x_{230} + 1 & x_{230}^2 + x_{231} \\
w_{131}^2 + w_{132} & w_{131}x_{131} + 1 & x_{131}^2 + x_{132} \\
w_{231}^2 + w_{232} & w_{231}x_{231} + 1 & x_{231}^2 + x_{232} \\
w_{132}^2 + w_{133} & w_{132}x_{132} + 1 & x_{132}^2 + x_{133} \\
w_{232}^2 + w_{233} & w_{232}x_{232} + 1 & x_{232}^2 + x_{233} \\
w_{133}^2 + w_{130} & w_{133}x_{133} + 1 & x_{133}^2 + x_{130} \\
w_{233}^2 + w_{230} & w_{233}x_{233} + 1 & x_{233}^2 + x_{230}
\end{array}
$$

Diffusion Relations: Round 1

$$w_{200} + Fx_{100} + 3x_{101} + 7x_{102} + Fx_{103} + Ax_{130} + 2x_{131} + Bx_{132} + Ax_{133} + k_{100} + 6$$
$$w_{201} + Ax_{100} + Ax_{101} + 5x_{102} + 6x_{103} + 8x_{130} + 8x_{131} + 4x_{132} + 9x_{133} + k_{101} + 7$$
$$w_{202} + 7x_{100} + 8x_{101} + 8x_{102} + 2x_{103} + Dx_{130} + Cx_{131} + Cx_{132} + 3x_{133} + k_{102} + 6$$
$$w_{203} + 4x_{100} + 6x_{101} + Cx_{102} + Cx_{103} + 5x_{130} + Ex_{131} + Fx_{132} + Fx_{133} + k_{103} + 7$$
$$w_{210} + Ax_{100} + 2x_{101} + Bx_{102} + Ax_{103} + Fx_{130} + 3x_{131} + 7x_{132} + Fx_{133} + k_{110} + 6$$
$$w_{211} + 8x_{100} + 8x_{101} + 4x_{102} + 9x_{103} + Ax_{130} + Ax_{131} + 5x_{132} + 6x_{133} + k_{111} + 7$$
$$w_{212} + Dx_{100} + Cx_{101} + Cx_{102} + 3x_{103} + 7x_{130} + 8x_{131} + 8x_{132} + 2x_{133} + k_{112} + 6$$
$$w_{213} + 5x_{100} + Ex_{101} + Fx_{102} + Fx_{103} + 4x_{130} + 6x_{131} + Cx_{132} + Cx_{133} + k_{113} + 7$$
$$w_{220} + Ax_{110} + 2x_{111} + Bx_{112} + Ax_{113} + Fx_{120} + 3x_{121} + 7x_{122} + Fx_{123} + k_{120} + 6$$
$$w_{221} + 8x_{110} + 8x_{111} + 4x_{112} + 9x_{113} + Ax_{120} + Ax_{121} + 5x_{122} + 6x_{123} + k_{121} + 7$$
$$w_{222} + Dx_{110} + Cx_{111} + Cx_{112} + 3x_{113} + 7x_{120} + 8x_{121} + 8x_{122} + 2x_{123} + k_{122} + 6$$
$$w_{223} + 5x_{110} + Ex_{111} + Fx_{112} + Fx_{113} + 4x_{120} + 6x_{121} + Cx_{122} + Cx_{123} + k_{123} + 7$$
$$w_{230} + Fx_{110} + 3x_{111} + 7x_{112} + Fx_{113} + Ax_{120} + 2x_{121} + Bx_{122} + Ax_{123} + k_{130} + 6$$
$$w_{231} + Ax_{110} + Ax_{111} + 5x_{112} + 6x_{113} + 8x_{120} + 8x_{121} + 4x_{122} + 9x_{123} + k_{131} + 7$$
$$w_{232} + 7x_{110} + 8x_{111} + 8x_{112} + 2x_{113} + Dx_{120} + Cx_{121} + Cx_{122} + 3x_{123} + k_{132} + 6$$
$$w_{233} + 4x_{110} + 6x_{111} + Cx_{112} + Cx_{113} + 5x_{120} + Ex_{121} + Fx_{122} + Fx_{123} + k_{133} + 7$$

Diffusion Relations: Round 2

$$c_{00} + Fx_{200} + 3x_{201} + 7x_{202} + Fx_{203} + Ax_{230} + 2x_{231} + Bx_{232} + Ax_{233} + k_{200} + 6$$
$$c_{01} + Ax_{200} + Ax_{201} + 5x_{202} + 6x_{203} + 8x_{230} + 8x_{231} + 4x_{232} + 9x_{233} + k_{201} + 7$$
$$c_{02} + 7x_{200} + 8x_{201} + 8x_{202} + 2x_{203} + Dx_{230} + Cx_{231} + Cx_{232} + 3x_{233} + k_{202} + 6$$
$$c_{03} + 4x_{200} + 6x_{201} + Cx_{202} + Cx_{203} + 5x_{230} + Ex_{231} + Fx_{232} + Fx_{233} + k_{203} + 7$$
$$c_{10} + Ax_{200} + 2x_{201} + Bx_{202} + Ax_{203} + Fx_{230} + 3x_{231} + 7x_{232} + Fx_{233} + k_{210} + 6$$
$$c_{11} + 8x_{200} + 8x_{201} + 4x_{202} + 9x_{203} + Ax_{230} + Ax_{231} + 5x_{232} + 6x_{233} + k_{211} + 7$$
$$c_{12} + Dx_{200} + Cx_{201} + Cx_{202} + 3x_{203} + 7x_{230} + 8x_{231} + 8x_{232} + 2x_{233} + k_{212} + 6$$
$$c_{13} + 5x_{200} + Ex_{201} + Fx_{202} + Fx_{203} + 4x_{230} + 6x_{231} + Cx_{232} + Cx_{233} + k_{213} + 7$$
$$c_{20} + Ax_{210} + 2x_{211} + Bx_{212} + Ax_{213} + Fx_{220} + 3x_{221} + 7x_{222} + Fx_{223} + k_{220} + 6$$
$$c_{21} + 8x_{210} + 8x_{211} + 4x_{212} + 9x_{213} + Ax_{220} + Ax_{221} + 5x_{222} + 6x_{223} + k_{221} + 7$$
$$c_{22} + Dx_{210} + Cx_{211} + Cx_{212} + 3x_{213} + 7x_{220} + 8x_{221} + 8x_{222} + 2x_{223} + k_{222} + 6$$
$$c_{23} + 5x_{210} + Ex_{211} + Fx_{212} + Fx_{213} + 4x_{220} + 6x_{221} + Cx_{222} + Cx_{223} + k_{223} + 7$$
$$c_{30} + Fx_{210} + 3x_{211} + 7x_{212} + Fx_{213} + Ax_{220} + 2x_{221} + Bx_{222} + Ax_{223} + k_{230} + 6$$
$$c_{31} + Ax_{210} + Ax_{211} + 5x_{212} + 6x_{213} + 8x_{220} + 8x_{221} + 4x_{222} + 9x_{223} + k_{231} + 7$$
$$c_{32} + 7x_{210} + 8x_{211} + 8x_{212} + 2x_{213} + Dx_{220} + Cx_{221} + Cx_{222} + 3x_{223} + k_{232} + 6$$
$$c_{33} + 4x_{210} + 6x_{211} + Cx_{212} + Cx_{213} + 5x_{220} + Ex_{221} + Fx_{222} + Fx_{223} + k_{233} + 7$$

Key Schedule Conjugacy Relations

$$
\begin{array}{lll}
k_{000}^2 + k_{001} & k_{100}^2 + k_{101} & k_{200}^2 + k_{201} \\
k_{001}^2 + k_{002} & k_{101}^2 + k_{102} & k_{201}^2 + k_{202} \\
k_{002}^2 + k_{003} & k_{102}^2 + k_{103} & k_{202}^2 + k_{203} \\
k_{003}^2 + k_{000} & k_{103}^2 + k_{100} & k_{203}^2 + k_{200} \\
k_{010}^2 + k_{011} & k_{110}^2 + k_{111} & k_{210}^2 + k_{211} \\
k_{011}^2 + k_{012} & k_{111}^2 + k_{112} & k_{211}^2 + k_{212} \\
k_{012}^2 + k_{013} & k_{112}^2 + k_{113} & k_{212}^2 + k_{213} \\
k_{013}^2 + k_{010} & k_{113}^2 + k_{110} & k_{213}^2 + k_{210} \\
k_{020}^2 + k_{021} & k_{120}^2 + k_{121} & k_{220}^2 + k_{221} \\
k_{021}^2 + k_{022} & k_{121}^2 + k_{122} & k_{221}^2 + k_{222} \\
k_{022}^2 + k_{023} & k_{122}^2 + k_{123} & k_{222}^2 + k_{223} \\
k_{023}^2 + k_{020} & k_{123}^2 + k_{120} & k_{223}^2 + k_{220} \\
k_{030}^2 + k_{031} & k_{130}^2 + k_{131} & k_{230}^2 + k_{231} \\
k_{031}^2 + k_{032} & k_{131}^2 + k_{132} & k_{231}^2 + k_{232} \\
k_{032}^2 + k_{033} & k_{132}^2 + k_{133} & k_{232}^2 + k_{233} \\
k_{033}^2 + k_{030} & k_{133}^2 + k_{130} & k_{233}^2 + k_{230}
\end{array}
$$

Key Schedule Inversion and Conjugacy Relations

$$
\begin{array}{llll}
k_{030}s_{000} + 1 & s_{000}^2 + s_{001} & k_{130}s_{100} + 1 & s_{100}^2 + s_{101} \\
k_{031}s_{001} + 1 & s_{001}^2 + s_{002} & k_{131}s_{101} + 1 & s_{101}^2 + s_{102} \\
k_{032}s_{002} + 1 & s_{002}^2 + s_{003} & k_{132}s_{102} + 1 & s_{102}^2 + s_{103} \\
k_{033}s_{003} + 1 & s_{003}^2 + s_{000} & k_{133}s_{103} + 1 & s_{103}^2 + s_{100} \\
k_{020}s_{010} + 1 & s_{010}^2 + s_{011} & k_{120}s_{110} + 1 & s_{110}^2 + s_{111} \\
k_{021}s_{011} + 1 & s_{011}^2 + s_{012} & k_{121}s_{111} + 1 & s_{111}^2 + s_{112} \\
k_{022}s_{012} + 1 & s_{012}^2 + s_{013} & k_{122}s_{112} + 1 & s_{112}^2 + s_{113} \\
k_{023}s_{013} + 1 & s_{013}^2 + s_{010} & k_{023}s_{113} + 1 & s_{113}^2 + s_{110}
\end{array}
$$

Key Schedule Diffusion Relations: Round 1

$$
\begin{array}{ll}
k_{100} + k_{000} & +5s_{000} + 1s_{001} + Cs_{002} + 5s_{003} + 7 \\
k_{101} + k_{001} & +2s_{000} + 2s_{001} + 1s_{002} + Fs_{003} + 6 \\
k_{102} + k_{002} & +As_{000} + 4s_{001} + 4s_{002} + 1s_{003} + 7 \\
k_{103} + k_{003} & +1s_{000} + 8s_{001} + 3s_{002} + 3s_{003} + 6 \\
k_{110} + k_{010} & +5s_{010} + 1s_{011} + Cs_{012} + 5s_{013} \\
k_{111} + k_{011} & +2s_{010} + 2s_{011} + 1s_{012} + Fs_{013} \\
k_{112} + k_{012} & +As_{010} + 4s_{011} + 4s_{012} + 1s_{013} \\
k_{113} + k_{013} & +1s_{010} + 8s_{011} + 3s_{012} + 3s_{013} \\
k_{120} + k_{020} + k_{000} & +5s_{000} + 1s_{001} + Cs_{002} + 5s_{003} + 7 \\
k_{121} + k_{021} + k_{001} & +2s_{000} + 2s_{001} + 1s_{002} + Fs_{003} + 6 \\
k_{122} + k_{022} + k_{002} & +As_{000} + 4s_{001} + 4s_{002} + 1s_{003} + 7 \\
k_{123} + k_{023} + k_{003} & +1s_{000} + 8s_{001} + 3s_{002} + 3s_{003} + 6 \\
k_{130} + k_{030} + k_{010} & +5s_{010} + 1s_{011} + Cs_{012} + 5s_{013} \\
k_{131} + k_{031} + k_{011} & +2s_{010} + 2s_{011} + 1s_{012} + Fs_{013} \\
k_{132} + k_{032} + k_{012} & +As_{010} + 4s_{011} + 4s_{012} + 1s_{013} \\
k_{133} + k_{033} + k_{013} & +1s_{010} + 8s_{011} + 3s_{012} + 3s_{013}
\end{array}
$$

Key Schedule Diffusion Relations: Round 2

$$
\begin{aligned}
k_{200} + k_{100} && +5s_{100} + 1s_{101} + Cs_{102} + 5s_{103} + 4 \\
k_{201} + k_{101} && +2s_{100} + 2s_{101} + 1s_{102} + Fs_{103} + 3 \\
k_{202} + k_{102} && +As_{100} + 4s_{101} + 4s_{102} + 1s_{103} + 5 \\
k_{203} + k_{103} && +1s_{100} + 8s_{101} + 3s_{102} + 3s_{103} + 2 \\
k_{210} + k_{110} && +5s_{110} + 1s_{111} + Cs_{112} + 5s_{113} \\
k_{211} + k_{111} && +2s_{110} + 2s_{111} + 1s_{112} + Fs_{113} \\
k_{212} + k_{112} && +As_{110} + 4s_{111} + 4s_{112} + 1s_{113} \\
k_{213} + k_{113} && +1s_{110} + 8s_{111} + 3s_{112} + 3s_{113} \\
k_{220} + k_{120} + k_{100} && +5s_{100} + 1s_{101} + Cs_{102} + 5s_{103} + 4 \\
k_{221} + k_{121} + k_{101} && +2s_{100} + 2s_{101} + 1s_{102} + Fs_{103} + 3 \\
k_{222} + k_{122} + k_{102} && +As_{100} + 4s_{101} + 4s_{102} + 1s_{103} + 5 \\
k_{223} + k_{123} + k_{103} && +1s_{100} + 8s_{101} + 3s_{102} + 3s_{103} + 2 \\
k_{230} + k_{130} + k_{110} && +5s_{110} + 1s_{111} + Cs_{112} + 5s_{113} \\
k_{231} + k_{131} + k_{111} && +2s_{110} + 2s_{111} + 1s_{112} + Fs_{113} \\
k_{232} + k_{132} + k_{112} && +As_{110} + 4s_{111} + 4s_{112} + 1s_{113} \\
k_{233} + k_{133} + k_{113} && +1s_{110} + 8s_{111} + 3s_{112} + 3s_{113}
\end{aligned}
$$

References

[1] K. Aoki and S. Vaudenay. On the Use of GF-Inversion as a Cryptographic Primitive. In M. Matsui and R. Zuccherato, editors, *Selected Areas in Cryptography (SAC) 2003*, volume 3006 of *LNCS*, pages 234–347. Springer–Verlag, 2004.

[2] F. Armknecht and S. Lucks. Linearity of the AES Key Schedule. In H. Dobbertin, V. Rijmen, and A. Sowa, editors, *Advanced Encryption Standard - AES, Fourth International Conference*, volume 3373 of *LNCS*, pages 145–162. Springer–Verlag, 2005.

[3] G. Ars. *Applications des Bases de Gröbner à la Cryptographie*. PhD thesis, Université de Rennes I, 2005.

[4] G. Ars, J-C. Faugère, H. Imai, M. Kawazoe, and M. Sugita. Comparison Between XL and Gröbner Basis Algorithms. In Pil Joong Lee, editor, *Advances in Cryptology - ASIACRYPT 2004*, volume 3329 of *LNCS*, pages 338–353. Springer–Verlag, 2004.

[5] M. Bardet. On the Complexity of a Gröbner Basis Algorithm. *Algorithms Seminar 2002-2004*, INRIA, 2005. http://algo.inria.fr/seminars.

[6] M. Bardet, J-C. Faugère, and B. Salvy. Complexity of Gröbner Basis Computation for Semi–Regular Overdetermined Sequences over F_2 with Solutions in F_2. *Technical Report* 5049, INRIA, 2003. http://www.inria.fr/rrrt/rr-5049.html.

[7] E. Barkan and E. Biham. In How Many Ways Can You Write Rijndael? In Y. Zheng, editor, *Advances in Cryptology - ASIACRYPT 2002*, volume 2501 of *LNCS*, pages 160–175. Springer–Verlag, 2002.

[8] E. Barkan and E. Biham. The Book of Rijndaels. *Cryptology ePrint Archive* 2002/158, 2002. http://eprint.iacr.org/2002/158/.

[9] E. Biham, R.J. Anderson, and L.R. Knudsen. SERPENT: A New Block Cipher Proposal. In S. Vaudenay, editor, *Fast Software Encryption 1998*, volume 1372 of *LNCS*, pages 222–238. Springer–Verlag, 1998.

[10] E. Biham and A. Shamir. Differential Cryptanalysis of DES-like Cryptosystems. *Journal of Cryptology*, 4:3–72, 1993.

[11] E. Biham and A. Shamir. *Differential Cryptanalysis of the Data Encryption Standard*. Springer–Verlag, 1993.

[12] A. Biryukov, C. De Cannière, A. Bracken, and B. Preneel. A Toolbox for Cryptanalysis: Linear and Affine Equivalence Algorithms. In E. Biham, editor, *Advances in Cryptology - EUROCRYPT 2003*, volume 2656 of *LNCS*, pages 33–50. Springer–Verlag, 2003.

[13] B. Buchberger. *Ein Algorithmus zum Auffinden der Basiselemente des Restklassenringes nach einem nulldimensionalen Polynomideal*. PhD thesis, Universität Innsbruck, 1965.

[14] J. Buchmann, A. Pychkine, and R-P. Weinmann. A Zero-dimensional Gröbner Basis for AES-128. In M.J.B. Robshaw, editor, *Fast Software Encryption 2006*, volume 4047 of *LNCS*. Springer–Verlag, 2006.

[15] K.W. Campbell and M.J. Wiener. DES is Not a Group. In E.F. Brickell, editor, *Advances in Cryptology - CRYPTO '92*, volume 740 of *LNCS*, pages 512–520. Springer–Verlag, 1993.

[16] D. Chaum and J-H. Evertse. Cryptanalysis of DES with a Reduced Number of Rounds. In H.C. Williams, editor, *Advances in Cryptology - CRYPTO '85*, volume 218 of *LNCS*, pages 192–211. Springer–Verlag, 1986.

[17] J-M. Chen, N. Courtois, and B-Y. Yang. On Asymptotic Security Estimates in XL and Gröbner Bases-Related Algebraic Cryptanalysis. In J. Lopez, S. Qing, and E. Okhamoto, editors, *ICICS*, volume 3269 of *LNCS*, pages 401–413. Springer–Verlag, 2004.

[18] J-M. Chen and B-Y. Yang. All in the XL Family: Theory and Practice. In C. Park and S. Chee, editors, *Proceedings of the 7th International Conference on Information Security and Cryptology*, volume 3506 of *LNCS*, pages 67–86. Springer–Verlag, 2004.

[19] J-M. Chen and B-Y. Yang. Theoretical Analysis of XL over Small Fields. In H. Wang, J. Piepryzk, and V. Varadharajan, editors, *Proceedings of the 9th Australasian Conference on Information Security and Privacy*, volume 3108 of *LNCS*, pages 277–288. Springer–Verlag, 2004.

[20] C. Cid and G. Leurent. An Analysis of the XSL Algorithm. In B. Roy, editor, *Advances in Cryptology - ASIACRYPT 2005*, volume 3788 of *LNCS*, pages 333–352. Springer–Verlag, 2005.

[21] C. Cid, S. Murphy, and M.J.B. Robshaw. An Algebraic Framework for Cipher Embeddings. In N.P. Smart, editor, *10th IMA International Conference on Coding and Cryptography*, volume 3796 of *LNCS*, pages 278–289. Springer–Verlag, 2005.

[22] C. Cid, S. Murphy, and M.J.B. Robshaw. Small Scale Variants of the AES. In H. Gilbert and H. Handschuh, editors, *Fast Software Encryption 2005*, volume 3557 of *LNCS*, pages 145–162. Springer–Verlag, 2005.

[23] P. Cohn. *Classical Algebra*. John Wiley, 2000.

[24] D. Coppersmith. The Real Reason for Rivest's Phenomenon. In H.C. Williams, editor, *Advances in Cryptology - CRYPTO '85*, volume 218 of *LNCS*, pages 535–536. Springer–Verlag, 1986.

[25] D. Coppersmith. The Data Encryption Standard (DES) and its Strength Against Attacks. Research Report RC 18613, IBM, 1992.

[26] N. Courtois. Algebraic Attacks over GF(2^k): Applications to HFE Challenge 2 and Sflash-v2. In F. Bao, R. Deng, and J. Zhou, editor, *Public Key Cryptography - PKC 2004*, volume 2947 of *LNCS*, pages 201–217. Springer–Verlag, 2004.

[27] N. Courtois. The Inverse S-Box, Non-linear Polynomial Relations and Cryptanalysis of Block Ciphers. In V. Rijmen H. Dobbertin and A. Sowa, editors, *Advanced Encryption Standard - AES, Fourth International Conference*, volume 3373 of *LNCS*, pages 234–347. Springer–Verlag, 2005.

[28] N. Courtois, A. Klimov, J. Patarin, and A. Shamir. Efficient Algorithms for Solving Overdefined Systems of Multivariate Polynomial Equations. In B. Preneel, editor, *Advances in Cryptology - EUROCRYPT 2000*, volume 1807 of *LNCS*, pages 392–407. Springer–Verlag, 2000.

[29] N. Courtois and W. Meier. Algebraic Attacks on Stream Ciphers with Linear Feedback. In E. Biham, editor, *Advances in Cryptology - EUROCRYPT 2003*, volume 2656 of *LNCS*, pages 345–359. Springer–Verlag, 2003.

[30] N. Courtois and J. Patarin. About the XL Algorithm over GF(2). In M. Joye, editor, *Progress in Cryptology - CT-RSA 2003*, pages 140–156. Springer–Verlag, 2003.

[31] N. Courtois and J. Pieprzyk. Cryptanalysis of Block Ciphers with Overdefined Systems of Equations. *Cryptology ePrint Archive* 2002/044, 2002. http://eprint.iacr.org/2002/044/.

[32] N. Courtois and J. Pieprzyk. Cryptanalysis of Block Ciphers with Overdefined Systems of Equations. In Y. Zheng, editor, *Advances in Cryptology - ASIACRYPT 2002*, volume 2501 of *LNCS*, pages 267–287. Springer–Verlag, 2002.

[33] D. Cox, J. Little, and D. O'Shea. *Ideals, Varieties, and Algorithms*. Undergraduate Texts in Mathematics. Springer–Verlag, second edition, 1997.

[34] D. Cox, J. Little, and D. O'Shea. *Using Algebraic Geometry*, volume 185 of *Graduate Texts in Mathematics*. Springer, second edition, 2004.

[35] J. Daemen. *Cipher and Hash Function Design Strategies based on Linear and Differential Cryptanalysis*. PhD thesis, Katholieke Universiteit Leuven, 1995.

[36] J. Daemen, L.R. Knudsen, and V. Rijmen. The Block Cipher SQUARE. In E. Biham, editor, *Fast Software Encryption 1997*, volume 1267 of *LNCS*, pages 149–165. Springer–Verlag, 1997.

[37] J. Daemen and V. Rijmen. Rijndael. *Submission to NIST AES Process*, 1997. http://csrc.nist.gov/CryptoToolkit/aes/.

[38] J. Daemen and V. Rijmen. Answer to "New Observations on Rijndael". *Submission to NIST AES Process*, 2000. http://csrc.nist.gov/CryptoToolkit/aes/.

[39] J. Daemen and V. Rijmen. *The Design of Rijndael*. Springer–Verlag, 2002.

[40] Y. Desmedt, T.V. Le, M. Marrotte, and J-J. Quisquater. Algebraic Structures and Cycling Test of Rijndael, 2001. http://citeseer.ist.psu.edu/desmedt02algebraic.html.

[41] C. Diem. The XL-Algorithm and a Conjecture from Commutative Algebra. In P.J. Lee, editor, *Advances in Cryptology – ASIACRYPT 2004*, volume 3329 of *LNCS*, pages 323–337. Springer–Verlag, 2004.

[42] H. Dobbertin, L.R. Knudsen, and M.J.B. Robshaw. The Cryptanalysis of the AES – A Brief Survey. In H. Dobbertin, V. Rijmen, and A. Sowa, editors, *Advanced Encryption Standard - AES, Fourth International Conference*, volume 3373 of *LNCS*, pages 1–10. Springer–Verlag, 2005.

[43] ECRYPT. The State of the Art of AES Cryptanalysis. Technical Report, ECRYPT Network of Excellence, 2005. http://www.ecrypt.eu.org.

[44] ECRYPT. AES Lounge. Website, ECRYPT Network of Excellence, 2006. http://www.iaik.tu-graz.ac.at/research/krypto/AES/.

[45] J-H. Evertse. Cryptanalysis of DES with a Reduced Number of Rounds. In D. Chaum and W.L. Price, editors, *Advances in Cryptology – EUROCRYPT 87*, volume 304 of *LNCS*, pages 249–266. Springer–Verlag, 1988.

[46] J-C. Faugère. A New Efficient Algorithm for Computing Gröbner bases (F4). *Journal of Pure and Applied Algebra*, 139:61–88, 1999.

[47] J-C. Faugère. A New Efficient Algorithm for Computing Gröbner Bases without Reduction to Zero (F5). In T. Mora, editor, *International Symposium on Symbolic and Algebraic Computation – ISSAC 2002*, pages 75–83, 2002.

[48] J-C. Faugère, P. Gianni, D. Lazard, and T. Mora. Efficient Computation of Zero-dimensional Gröbner Bases by Change of Ordering. *Journal of Symbolic Computation*, 16(4):329–344, 1993.

[49] J-C. Faugère and A. Joux. Algebraic Cryptanalysis of Hidden Field Equation (HFE) Cryptosystems using Gröbner Bases. In D. Boneh, editor, *Advances in Cryptology – CRYPTO 2003*, volume 2729 of *LNCS*, pages 44–60. Springer–Verlag, 2003.

[50] H. Feistel. Cryptography and Computer Piracy. *Scientific American*, 228:15–23, 1973.

[51] N. Ferguson, J. Kelsey, B. Schneier, M. Stay, D. Wagner, and D. Whiting. Improved Cryptanalysis of Rijndael. In B. Schneier, editor, *Fast Software Encryption 2000*, volume 1978 of *LNCS*, pages 213–230. Springer–Verlag, 2000.

[52] N. Ferguson, R. Schroeppel, and D. Whiting. A Simple Algebraic Representation of Rijndael. In S. Vaudenay and A. Youssef, editors, *Selected Areas in Cryptography (SAC) 2001*, volume 2259 of *LNCS*, pages 103–111. Springer–Verlag, 2001.

[53] J. Fuller and W. Millan. Linear Redundancy in S-Boxes. In T. Johansson, editor, *Fast Software Encryption 2003*, volume 2887 of *LNCS*, pages 74–86. Springer–Verlag, 2003.

[54] M.R. Garey and D.S. Johnson. *Computers and Intractability - A Guide to the Theory of NP-Completeness.* W.H. Freeman and Company, 1979.

[55] H. Gilbert and M. Minier. A Collision Attack on 7 Rounds of Rijndael. In *Proceedings of Third Advanced Encryption Standard Conference*, pages 230–241. National Institute of Standards and Technology, 2000.

[56] M. Hellman, R. Merkle, R. Schroeppel, L. Washington, W. Diffie, S. Pohlig, and P. Schweitzer. Results of an Initial Attempt to Cryptanalyse the NBS Data Encryption Standard. *Technical Report* 76–042, Stanford University Electronics Laboratories, 1976.

[57] I.N. Herstein. *Topics in Algebra.* John Wiley & Sons, Second edition, 1975.

[58] J.W.P. Hirschfeld. *Projective Geometry over Finite Fields.* Oxford Mathematical Monographs. Oxford University Press, 1998.

[59] K. Hoffman and R. Kunze. *Linear Algebra.* Prentice-Hall, Second edition, 1971.

[60] G. Hornauer, W. Stephan, and R. Wernsdorf. Markov Ciphers and Alternating Groups. In T. Helleseth, editor, *Advances in Cryptology - EUROCRYPT '93*, volume 765 of *LNCS*, pages 453–460. Springer–Verlag, 1994.

[61] W-A. Jackson and S. Murphy. Projective Aspects of the AES Inversion. *Technical Report* RHUL–MA–2006–4, Royal Holloway, University of London, 2005. http://www.ma.rhul.ac.uk/techreports/.

[62] T. Jakobsen and L. Knudsen. Attacks on Block Ciphers of Low Algebraic Degree. *Journal of Cryptology*, 14:197–210, 2001.

[63] T. Jakobsen and L.R. Knudsen. The Interpolation Attack on Block Ciphers. In E. Biham, editor, *Fast Software Encryption 1997*, volume 1267 of *LNCS*, pages 28–40. Springer–Verlag, 1997.

[64] D. Kahn. *Seizing the Enigma.* Arrow Books, 1996.

[65] B.S. Kaliski, R. Rivest, and A.T. Sherman. Is the Data Encryption Standard a Group? (Results of Cycling Experiments on DES). *Journal of Cryptology*, 1:3–36, 1988.

[66] M. Kalkbrener. Solving Systems of Algebraic Equations by Using Gröbner Bases. In J.H. Davenport, editor, *EUROCAL'87*, volume 378 of *LNCS*, pages 282–291. Springer–Verlag, 1989.

[67] L. Knudsen and H. Raddum. Recommendation to NIST for the AES. *Submission to NIST AES Process*, 2000. http://csrc.nist.gov/CryptoToolkit/aes/.

[68] X. Lai, J.L. Massey, and S. Murphy. Markov Ciphers and Differential Cryptanalysis. In D.W. Davies, editor, *Advances in Cryptology - EUROCRYPT 91*, volume 547 of *LNCS*, pages 17–38. Springer–Verlag, 1991.

[69] S. Landau. Communications Security for the Twenty-First Century: The Advanced Encryption Standard. *Notices of the American Mathematical Society*, 47:450–459, 2000.

[70] S. Landau. Standing the Test of Time: The Data Encryption Standard. *Notices of the American Mathematical Society*, 47:341–349, 2000.

[71] S. Landau. Polynomials in the Nation's Service: Using Algebra to Design the Advanced Encryption Standard. *American Mathematical Monthly*, 111:89–117, 2004.

[72] D. Lazard. Gröbner Bases, Gaussian Elimination and Resolution of Systems of Algebraic Equations. In J.A. van Hulzen, editor, *Proceedings of the European Computer Algebra Conference on Computer Algebra*, volume 162 of *LNCS*, pages 146–156. Springer–Verlag, 1983.

[73] T. Van Lee, R. Sparr, R. Wernsdorf, and Y. Desmedt. Complementation–like and Cyclic Properties of AES Round Functions. In H. Dobbertin, V. Rijmen, and A. Sowa, editors, *Advanced Encryption Standard - AES, Fourth International Conference*, volume 3373 of *LNCS*, pages 128–141. Springer–Verlag, 2005.

[74] R. Lidl and H. Niederreiter. *Introduction to Finite Fields and their Applications*. Cambridge University Press, Revised edition, 1994.

[75] D. Mackenzie. A Game of Chance. *New Scientist*, 2398:36–39, 2003. 7 June 2003.

[76] F.J. MacWilliams and N.J.A. Sloane. *The Theory of Error-Correcting Codes*. North-Holland Publishing Company, 1978.

[77] MAGMA v2.12. Computational Algebra Group, School of Mathematics and Statistics, University of Sydney, 2005. http://magma.maths.usyd.edu.au.

[78] M. Matsui. Linear Cryptanalysis Method for DES Cipher. In T. Helleseth, editor, *Advances in Cryptology - EUROCRYPT '93*, volume 765 of *LNCS*, pages 386–397. Springer–Verlag, 1994.

[79] E.W. Mayr. Some Complexity Results for Polynomial Ideals. *Journal of Complexity*, 13(3):303–325, 1997.

[80] A.J. Menezes, P.C. Van Oorschot, and S.A. Vanstone. *Handbook of Applied Cryptography*. CRC Press, 1996.

[81] T. Moh. On the Method of XL and its Inefficiency against TTM. *Cryptology ePrint Archive* 2001/047, 2001. http://eprint.iacr.org/2001/047/.

[82] J. Monnerat and S. Vaudenay. On some Weak Extensions of AES and BES. In J. Lopez, S. Qinq, and E. Okhamoto, editors, *Sixth International Conference on Information and Communications Security*, volume 3269 of *LNCS*, pages 414–426. Springer–Verlag, 2004.

[83] J.H. Moore and G.J. Simmons. Cycle Structures of the DES with Weak and Semi-Weak Keys. In A.M. Odlyzko, editor, *Advances in Cryptology - CRYPTO '86*, volume 263 of *LNCS*, pages 9–32. Springer–Verlag, 1987.

[84] S. Murphy. An Analysis of SAFER. *Journal of Cryptology*, 11:235–251, 1998.

[85] S. Murphy, K.G. Paterson, and P. Wild. A Weak Cipher that Generates the Symmetric Group. *Journal of Cryptology*, 7:61–65, 1994.

[86] S. Murphy, F. Piper, M. Walker, and P. Wild. Maximum Likelihood Estimation for Block Cipher Keys. *Technical Report* RHUL–MA–2006-3, Royal Holloway, University of London, 1994. http://www.ma.rhul.ac.uk/techreports/.

[87] S. Murphy and M.J.B. Robshaw. Further Comments on the Structure of Rijndael. *Submission to NIST AES Process*, 2000. http://csrc.nist.gov/CryptoToolkit/aes/.

[88] S. Murphy and M.J.B. Robshaw. New Observations on Rijndael. *Submission to NIST AES Process*, 2000. http://csrc.nist.gov/CryptoToolkit/aes/.

[89] S. Murphy and M.J.B. Robshaw. Essential Algebraic Structure Within the AES. In M. Yung, editor, *Advances in Cryptology - CRYPTO 2002*, volume 2442 of *LNCS*, pages 1–16. Springer–Verlag, 2002.

[90] S. Murphy and M.J.B. Robshaw. Comments on the Security of the AES and the XSL Technique. *Electronic Letters*, 39:26–38, 2003.

[91] M.A. Musa, E.F. Schaefer, and S. Wedig. A Simplified AES Algorithm and its Linear and Differential Cryptanalysis. *Cryptologia*, XXVII (2):148–177, 2003.

[92] National Bureau of Standards. The Data Encryption Standard. Federal Information Processing Standards Publication (FIPS) 46, 1977.

[93] National Institute of Standards and Technology. The Digital Signature Standard. Federal Information Processing Standards Publication (FIPS) 186, 1994.

[94] National Institute of Standards and Technology. Announcing the Development of a Federal Information Processing Standard for Advanced Encryption Standard, 1997.

[95] National Institute of Standards and Technology. The Advanced Encryption Standard. Federal Information Processing Standards Publication (FIPS) 197, 2001.

[96] National Institute of Standards and Technology. Recommendation for the Triple Data Encryption Algorithm (TDEA) Block Cipher. Special Publication SP800-67, 2004.

[97] C.W. Norman. *Undergraduate Algebra*. Oxford University Press, 1986.

[98] K. Nyberg. Differentially Uniform Mappings for Cryptography. In T. Helle-seth, editor, *Advances in Cryptology - EUROCRYPT '93*, volume 765 of *LNCS*, pages 55–64. Springer–Verlag, 1994.

[99] K. Nyberg and L.R. Knudsen. Provable Security Against a Differential Attack. *Journal of Cryptology*, 8(1):27–38, 1995.

[100] J. Patarin. Hidden Fields Equations (HFE) and Isomorphisms of Polynomials (IP): Two New Families of Asymmetric Algorithms. In U. Maurer, editor, *Advances in Cryptology - EUROCRYPT '96*, volume 1070 of *LNCS*, pages 33–48. Spinger–Verlag, 1996.

[101] K.G. Paterson. Imprimitive Permutation Groups and Trapdoors in Iterated Block Ciphers. In L.R. Knudsen, editor, *Fast Software Encryption 1999*, volume 1636 of *LNCS*, pages 201–214. Springer–Verlag, 1999.

[102] R.C.-W. Phan. Mini Advanced Encryption Standard (Mini-AES): A Testbed for Cryptanalysis Students. *Cryptologia*, XXVI (4):283–306, 2002.

[103] J-J. Quisquater and J-P. Delescaille. How Easy is Collision Search? Application to DES. In J-J. Quisquater and J. Vandewalle, editors, *Advances in Cryptology - EUROCRYPT '89*, volume 434 of *LNCS*, pages 429–434. Springer–Verlag, 1990.

[104] J-J. Quisquater and J-P. Delescaille. How Easy is Collision Search? New Results and Applications to DES. In G. Brassard, editor, *Advances in Cryptology - CRYPTO '89*, volume 435 of *LNCS*, pages 408–413. Springer–Verlag, 1990.

[105] H. Raddum. More Dual Rijndaels. In H. Dobbertin, V. Rijmen, and A. Sowa, editors, *Advanced Encryption Standard - AES, Fourth International Conference*, volume 3373 of *LNCS*, pages 142–147. Springer–Verlag, 2005.

[106] J.A. Reeds and J.L. Manfredelli. DES has no Per Round Linear Factors. In G.R Blakely and D. Chaum, editors, *Advances in Cryptology - Proceedings of CRYPTO 84*, volume 196 of *LNCS*, pages 377–389. Springer–Verlag, 1985.

[107] V. Rijmen. *Cryptanalysis and Design of Iterated Block Ciphers*. PhD thesis, Katholieke Universiteit Leuven, 1997.

[108] V. Rijmen, J. Daemen, B. Preneel, A. Bosselaers, and E. De Win. The Cipher SHARK. In D. Gollman, editor, *Fast Software Encryption 1996*, volume 1039 of *LNCS*, pages 99–112. Springer–Verlag, 1996.

[109] I. Schaumueller-Bichl. Cryptanalysis of the Data Encryption Standard by the Method of Formal Coding. In T. Beth, editor, *Proceedings of Workshop on Cryptography, Berg Fuerstein, Germany 1982 (EUROCRYPT 82)*, volume 143 of *LNCS*, pages 235–255. Springer–Verlag, 1983.

[110] B. Schneier. AES News. *Crypto-Gram Newsletter*, September 2002. http://www.schneier.com/crypto-gram-0209.html.

[111] R. Schroeppel. Second Round Comments to NIST. *Submission to NIST AES Process*, 1999. http://csrc.nist.gov/CryptoToolkit/aes/.

[112] C. Seife. Crucial Cipher Flawed, Cryptographers Claim. *Science*, 297:2193, 2002. 27 September 2002.

[113] C.E. Shannon. Communication Theory of Secrecy Systems. *Bell System Technical Journal*, 28-4:656–715, 1949.

[114] T. Shimoyama and T. Kaneko. Quadratic Relation of S-box and Its Application to the Linear Attack of Full Round DES. In H. Krawczyk, editor, *Advances in Cryptology – CRYPTO '98*, volume 1462 of *LNCS*, pages 200–211. Springer–Verlag, 1998.

[115] A. Steel. Computing Gröbner Bases with Linear Algebra. *Algebraic Geometry and Number Theory with* Magma, Institute Henri Poincaré, Paris, 2004.

[116] I. Toli and A. Zanoni. An Algebraic Interpretation of AES-128. In H. Dobbertin, V. Rijmen, and A. Sowa, editors, *Advanced Encryption Standard - AES, Fourth International Conference*, volume 3373 of *LNCS*, pages 84–97. Springer–Verlag, 2005.

[117] D. Wagner. Towards a Unifying View of Block Cipher Cryptanalysis. In B. Roy and W. Meier, editors, *Fast Software Encryption 2004*, volume 3017 of *LNCS*, pages 16–33. Springer–Verlag, 2004.

[118] R. Wernsdorf. The One-Round Functions of the DES Generate the Alternating Group. In R.A. Rueppel, editor, *Advances in Cryptology – EUROCRYPT 1992*, volume 658 of *LNCS*, pages 99–112. Springer–Verlag, 1993.

[119] R. Wernsdorf. The Round Functions of RIJNDAEL Generate the Alternating Group. In J. Daemen and V. Rijmen, editors, *Fast Software Encryption 2002*, volume 2365 of *LNCS*, pages 143–148. Springer–Verlag, 2002.

[120] A.M. Youssef and S.E. Tavares. Affine Equivalence in the AES Round Function. *Discrete Applied Mathematics*, 148(2):161–170, 2005.

Index

Carlos Cid has a B.Sc. and a Ph.D. in Mathematics from the University of Brasilia. He is currently a lecturer in the Information Security Group at Royal Holloway (University of London). He has worked as a security engineer in Ireland and has held academic positions as a lecturer at the University of Brasilia and a postdoctoral researcher at RWTH-Aachen.

Sean Murphy has a B.A. in Mathematics from the University of Oxford and a Ph.D. in Mathematics from the University of Bath. He is currently a Professor in the Information Security Group at Royal Holloway (University of London). He was a member of the European *NESSIE* project for evaluating cryptographic standards and of the executive committee of *ECRYPT*, the European Network of Excellence in Cryptology. He is a co-author of *Cryptography: A Very Short Introduction*.

Matthew Robshaw has a B.Sc. in Mathematics from St. Andrews University and a Ph.D. in Mathematics from Royal Holloway (University of London). He is currently a cryptographic researcher at France Telecom R&D. He was a Reader in the Information Security Group at Royal Holloway (University of London). Prior to that, he was a Principal Research Scientist at RSA Laboratories (California), where he was a co-designer of the AES finalist RC6.